Financing Green Infrastructure: Innovative Strategies for Sustainable Urban Water Management and Climate Resilience

Copyright

Financing Green Infrastructure: Innovative Strategies for Sustainable Urban Water Management and Climate Resilience

© 2025 Robert C. Brears

Author: Robert C. Brears

Published by:

Global Climate Solutions

ISBNs:

978-1-991369-03-1 (eBook)

978-1-991369-04-8 (Paperback)

The information presented in this book is for educational and informational purposes only. The author has made every effort to ensure accuracy but assumes no responsibility for errors or omissions.

Table of Contents

Preface

In an era of rapid urbanization and escalating climate challenges, cities worldwide are grappling with the dual pressures of managing water resources and safeguarding against environmental risks. Traditional water infrastructure—characterized by gray systems like concrete stormwater drains and levees—has proven inadequate in addressing the complexity of modern urban water challenges. The need for a paradigm shift toward sustainable solutions has never been more pressing.

Green infrastructure offers a transformative approach to urban water management, combining nature-based solutions like rain gardens, green roofs, and permeable pavements with ecological resilience and cost efficiency. Unlike conventional systems, green infrastructure not only addresses immediate water management needs but also provides co-benefits, including enhanced urban biodiversity, improved air quality, and vibrant community spaces. Yet, despite its benefits, funding remains a significant barrier to its widespread adoption.

This book explores how cities are securing the necessary financial resources to implement and scale green infrastructure projects. From Environmental Impact Bonds (EIBs) to public funding mechanisms, innovative financial models are enabling solutions that tackle stormwater management, flood risk reduction, and urban resilience. These funding strategies are reshaping the way urban areas approach sustainability, blending performance-based investments with forward-thinking governance.

Through this guide, you will uncover practical insights into how cities, governments, and the private sector are collaborating to unlock the potential of green infrastructure. The chapters ahead delve into the mechanisms, tools, and strategies that are empowering urban centers to navigate the complexities of financing sustainable solutions. By understanding these approaches, stakeholders can overcome financial barriers and create scalable, impactful projects that benefit both the environment and urban communities.

As we explore the evolving landscape of green infrastructure financing, this book serves as a blueprint for action—equipping decision-makers, policymakers, and industry leaders with the knowledge to implement innovative and sustainable water management strategies. Together, we can transform urban spaces into resilient, climate-adaptive cities for future generations.

Introduction

Cities around the world are facing increasing challenges in managing water resources due to rapid urbanization and the growing impacts of climate change. Traditional gray infrastructure, such as concrete drainage systems, is proving insufficient to address complex issues like stormwater management, flood prevention, and urban water scarcity. In response, green infrastructure has emerged as a sustainable alternative, integrating nature-based solutions that offer environmental, social, and economic benefits. However, a significant barrier to its widespread adoption is the availability of adequate funding. This book examines innovative financial mechanisms that enable the implementation and scaling of green infrastructure to build resilient urban environments.

Green infrastructure plays a pivotal role in addressing the challenges of urban water management, offering sustainable solutions to issues that conventional systems cannot fully resolve. Unlike traditional gray infrastructure, green infrastructure incorporates nature-based solutions such as rain gardens, green roofs, and permeable pavements, which work with natural processes to manage stormwater, reduce flood risks, and improve water quality. These systems not only address functional water management needs but also provide significant environmental, social, and economic co-benefits.

By enhancing urban biodiversity, improving air quality, and creating attractive, multifunctional spaces, green infrastructure fosters healthier, more livable cities. Its adaptability to climate change impacts, such as increasing rainfall intensity and prolonged droughts, makes it a vital tool for building urban resilience. However, despite its clear advantages, the implementation of green infrastructure remains limited, largely due to financial constraints. Unlocking innovative funding mechanisms is essential to scale these solutions and ensure sustainable urban growth.

Financing plays a crucial role in advancing sustainability by enabling the implementation of innovative solutions such as green infrastructure. Sustainable development often requires substantial upfront investment, which can be a barrier for many cities and communities, especially those with limited financial resources. Effective financing models help bridge this gap, allowing urban areas to transition from conventional infrastructure to nature-based solutions that address pressing water management challenges.

Innovative mechanisms, such as EIBs, green bonds, and public-private partnerships, align funding with measurable environmental and social outcomes, encouraging investment in resilient urban systems. These models provide a framework for allocating resources efficiently, minimizing risks, and achieving long-term benefits. Moreover, financing not only supports project implementation but also drives innovation by incentivizing stakeholders to develop scalable and cost-effective approaches. By prioritizing financial strategies tailored to sustainability, cities can overcome funding barriers and create resilient, climate-adaptive urban spaces for future generations.

This book aims to provide a comprehensive guide to financing green infrastructure as a solution to urban water management challenges. It explores the innovative financial mechanisms and strategies that cities, governments, and private entities are leveraging to implement sustainable water management systems. The chapters examine tools such as EIBs, public funding, and private investments, offering insights into their structure, benefits, and challenges. The book also addresses barriers to financing, the integration of green infrastructure into urban planning, and the role of technology in optimizing funding processes. Together, these topics create a blueprint for building resilient, sustainable cities.

Chapter 1: The Need for Green Infrastructure in Urban Water Management

This chapter explores why green infrastructure is essential for urban water management in the 21st century. It begins by defining green infrastructure and its core principles, emphasizing its adaptability to diverse urban contexts. The discussion then highlights the limitations of gray infrastructure, the mounting challenges cities face, and the growing recognition of green solutions as a more sustainable alternative. Finally, the chapter delves into the co-benefits of green infrastructure, demonstrating how it aligns with global sustainability goals while addressing local needs.

By understanding the pressing need for green infrastructure, stakeholders can better appreciate its role in creating climate-resilient, vibrant urban environments that balance ecological health and human development. This foundational discussion sets the stage for examining innovative financial models that enable the implementation and scaling of these solutions in subsequent chapters.

Defining Green Infrastructure

Green infrastructure refers to a network of natural and semi-natural systems designed to manage water resources sustainably while providing additional environmental, social, and economic benefits. Unlike traditional gray infrastructure, which relies on engineered systems like pipes and concrete channels to address water management, green infrastructure works in harmony with natural processes to deliver multiple outcomes. It encompasses a wide range of practices, including rain gardens, green roofs, permeable pavements, bioswales, urban wetlands, and tree canopies.

At its core, green infrastructure focuses on mimicking and enhancing the water cycle by capturing, storing, and filtering stormwater where it falls. This approach reduces runoff, mitigates flooding, and improves water quality by allowing natural infiltration and filtration. By using vegetation and soil systems, green infrastructure also provides habitat for urban wildlife, contributes to carbon sequestration, and reduces the urban heat island effect. These characteristics make it a vital tool for addressing modern urban challenges, particularly in the face of climate change.

A key distinction between green and gray infrastructure lies in their flexibility and co-benefits. While gray infrastructure typically addresses a single function, such as diverting stormwater, green infrastructure achieves multiple objectives simultaneously. For instance, a green roof not only manages stormwater but also reduces building energy use, improves air quality, and offers recreational space. This multifunctionality ensures that investments in green infrastructure deliver broader societal value, making it an attractive alternative for urban planners and policymakers.

Green infrastructure is inherently scalable and adaptable, suitable for deployment in various urban contexts. From individual buildings to neighborhood and city-wide applications, it can be customized to meet specific needs and priorities. For example, permeable pavements can be used in parking lots to reduce runoff, while bioswales can be integrated into road medians to capture and filter stormwater. This adaptability allows cities to incorporate green infrastructure into existing urban systems, gradually transitioning from traditional approaches to sustainable alternatives.

Another defining feature of green infrastructure is its alignment with sustainability goals. It contributes to several United Nations Sustainable Development Goals (SDGs), including clean water and sanitation (SDG 6), sustainable cities and communities (SDG 11), and climate action (SDG 13). By addressing water management challenges in an environmentally and socially responsible manner, green infrastructure supports long-term urban resilience.

Despite its clear advantages, the implementation of green infrastructure faces barriers, such as funding constraints, regulatory hurdles, and limited public awareness. Overcoming these challenges requires collaboration among governments, private stakeholders, and communities to design, fund, and maintain green systems effectively.

In summary, green infrastructure represents a forward-thinking approach to urban water management, integrating natural processes with built environments to achieve sustainable outcomes. Its ability to address multiple urban challenges simultaneously makes it a cornerstone of modern city planning, offering a path toward resilient, equitable, and environmentally friendly urban landscapes. This chapter lays the groundwork for understanding why green infrastructure is essential in addressing the water management challenges cities face today.

Challenges of Traditional Infrastructure

Traditional gray infrastructure has long been the backbone of urban water management systems, including stormwater drains, sewers, levees, and dams. These engineered solutions are designed to channel water quickly and efficiently out of urban areas, protecting against flooding and ensuring reliable water supply. However, as cities face growing environmental, social, and economic pressures, the limitations of gray infrastructure are becoming increasingly evident.

One of the primary challenges of traditional infrastructure is its inflexibility in responding to changing climatic conditions. Systems designed for historical weather patterns often fail to cope with the increasing frequency and intensity of storms caused by climate change. Overwhelmed stormwater systems result in flooding, property damage, and disruptions to daily life. Similarly, prolonged droughts can strain water supply systems, exposing the vulnerabilities of infrastructure that relies heavily on predictable water cycles.

Another significant drawback of gray infrastructure is its high maintenance and replacement costs. Aging infrastructure in many cities is deteriorating, leading to frequent repairs and failures. For example, corroded pipes, cracked stormwater channels, and outdated treatment facilities require costly upgrades. This financial burden is particularly acute in cities with limited budgets, where maintaining existing systems competes with the need for new investments in resilience and sustainability.

Gray infrastructure also lacks the multifunctionality of modern solutions. Its primary focus on a single task—such as diverting stormwater or supplying water—means it often overlooks broader environmental and social benefits. This single-purpose design can exacerbate ecological degradation. For instance, channelized rivers and concrete drains disrupt natural water flows, harm aquatic ecosystems, and contribute to water pollution by quickly funneling untreated runoff into waterways.

Environmental degradation is another pressing concern associated with traditional infrastructure. Rapid urbanization has led to the widespread replacement of natural landscapes with impervious surfaces like asphalt and concrete. These surfaces prevent water from infiltrating the ground, increasing surface runoff and reducing groundwater recharge. The result is a cycle of urban water challenges: reduced water quality, depleted aquifers, and increased flood risks.

Socially, gray infrastructure can perpetuate inequities in urban areas. In many cases, low-income communities bear the brunt of failing infrastructure, experiencing higher rates of flooding and water contamination. These communities are often located in areas with outdated or poorly maintained systems, compounding their vulnerability to water-related risks. Furthermore, traditional infrastructure tends to prioritize technical efficiency over community involvement, limiting opportunities for public engagement and co-benefits like improved urban spaces.

Traditional infrastructure is also energy-intensive, contributing to greenhouse gas emissions and increasing cities' carbon footprints. For example, water treatment facilities, pumping stations, and desalination plants require significant amounts of energy to operate. As urban areas expand, the energy demands of gray infrastructure grow, creating a feedback loop of environmental harm.

Finally, the limited adaptability of gray infrastructure poses a long-term challenge. These systems are often designed with fixed capacities and cannot easily accommodate changes in population density or land use. For example, urban sprawl places additional strain on stormwater systems, leading to higher risks of flooding and water quality degradation. Retrofitting or expanding these systems is often costly and disruptive, requiring extensive planning and resources.

In contrast to these limitations, green infrastructure offers a more resilient and sustainable alternative. It provides flexible, cost-effective solutions that adapt to changing conditions while delivering multiple co-benefits, such as improving air quality, enhancing biodiversity, and creating attractive public spaces. However, transitioning to green infrastructure requires a clear understanding of the shortcomings of gray systems and a commitment to addressing them through innovative planning and financing.

In summary, the challenges of traditional infrastructure highlight its inadequacy in addressing the complex water management needs of modern cities. Climate change, urbanization, and aging systems have exposed its vulnerabilities, creating an urgent need for more sustainable and adaptable approaches. By understanding these limitations, cities can move toward green infrastructure solutions that balance environmental, social, and economic priorities, ensuring long-term resilience and sustainability.

Benefits of Green Solutions

Green infrastructure offers numerous benefits that address the limitations of traditional gray infrastructure while contributing to more sustainable and resilient urban environments. By integrating nature-based systems into urban planning, green solutions provide a multifaceted approach to managing water, improving environmental quality, and enhancing community well-being.

Effective Stormwater Management

One of the most prominent benefits of green infrastructure is its ability to manage stormwater effectively. Unlike traditional systems that rapidly channel runoff into waterways, green solutions like rain gardens, permeable pavements, and bioswales capture, filter, and store water at its source. These systems reduce the volume and velocity of stormwater runoff, mitigating flood risks and minimizing erosion. By allowing water to infiltrate the ground, they also replenish groundwater supplies, addressing water scarcity concerns.

Improved Water Quality

Green infrastructure naturally filters pollutants from stormwater, improving the quality of water entering streams, rivers, and lakes. Vegetation and soil systems trap sediments, nutrients, and contaminants, preventing them from reaching aquatic ecosystems. This function reduces the need for energy-intensive water treatment processes, lowering costs and carbon footprints. Cleaner water also supports healthier ecosystems and provides safer recreational opportunities for urban communities.

Enhanced Urban Resilience

As cities face more frequent and intense weather events due to climate change, green solutions contribute to urban resilience by mitigating the impacts of extreme conditions. Green roofs and urban tree canopies reduce heat absorption, lowering temperatures in densely built areas and combating the urban heat island effect. These systems also provide insulation for buildings, reducing energy consumption and costs during hot and cold weather. By addressing climate-related challenges holistically, green infrastructure makes cities more adaptable to future uncertainties.

Biodiversity and Habitat Creation

Green infrastructure fosters biodiversity by creating habitats for plants, birds, insects, and other wildlife in urban settings. Restored wetlands, green corridors, and urban forests provide critical refuges for native species, supporting ecological balance. These systems also serve as migration pathways, allowing wildlife to thrive even in highly developed areas. Enhanced biodiversity contributes to ecosystem services, such as pollination and pest control, which are essential for sustainable urban living.

Economic Advantages

Investing in green solutions can generate significant economic benefits. By reducing flood damage, improving water quality, and enhancing energy efficiency, green infrastructure lowers long-term costs for municipalities and property owners. In addition, green spaces increase property values, attract businesses, and promote tourism by creating attractive urban environments. Job creation is another key advantage, as designing, constructing, and maintaining green infrastructure projects provide opportunities for local employment.

Social and Health Benefits

Green infrastructure improves the quality of life for urban residents by creating accessible, aesthetically pleasing public spaces. Parks, greenways, and community gardens promote physical activity, mental well-being, and social interaction. Exposure to green spaces has been shown to reduce stress, enhance cognitive function, and improve overall health outcomes. Furthermore, by reducing air pollution and mitigating heat, green infrastructure contributes to healthier urban environments, particularly for vulnerable populations.

Contribution to Climate Goals

Green infrastructure aligns with global climate goals by reducing greenhouse gas emissions and enhancing carbon sequestration. Urban forests, wetlands, and vegetated systems capture and store carbon dioxide, helping cities meet emissions reduction targets. Additionally, the energy savings from green roofs and shaded urban areas lower the reliance on fossil fuels, further contributing to climate mitigation efforts.

Multifunctionality and Adaptability

Unlike gray infrastructure, which typically serves a single purpose, green solutions are multifunctional. For instance, a bioswale not only manages stormwater but also provides aesthetic value, wildlife habitat, and pollutant filtration. Green infrastructure's adaptability allows it to be integrated into existing urban landscapes, ranging from individual buildings to citywide networks. This flexibility makes it suitable for diverse urban contexts, from densely populated metropolitan areas to smaller communities.

Equity and Community Engagement

Green solutions offer opportunities to address social equity by providing underserved communities with improved access to green spaces and cleaner environments. Community-based projects empower residents to participate in the planning and maintenance of green infrastructure, fostering a sense of ownership and inclusion. By prioritizing equity, cities can ensure that the benefits of green infrastructure are distributed fairly across all neighborhoods.

Link to Urban Resilience and Water Challenges

Urban resilience refers to a city's ability to anticipate, prepare for, respond to, and recover from various shocks and stresses, including those caused by climate change, population growth, and resource scarcity. Water challenges are among the most significant threats to urban resilience, as they affect critical systems like stormwater management, potable water supply, and flood prevention. Green infrastructure offers a pathway to addressing these challenges while enhancing the overall resilience of urban areas.

Managing Increasing Stormwater Runoff

One of the most pressing water challenges in urban areas is the management of stormwater. As cities grow, impervious surfaces like roads, buildings, and parking lots replace natural landscapes, preventing water from infiltrating the ground. This leads to increased stormwater runoff, which can overwhelm drainage systems, cause localized flooding, and pollute waterways. Green infrastructure helps mitigate these impacts by capturing, storing, and treating stormwater where it falls. Systems such as bioswales, rain gardens, and permeable pavements allow water to infiltrate the soil, reducing runoff volume and preventing overflows in sewer systems.

Addressing Flood Risks

Flooding is another critical challenge that undermines urban resilience. Extreme weather events, exacerbated by climate change, are increasing in frequency and intensity, placing significant strain on urban infrastructure. Traditional gray infrastructure, such as levees and stormwater drains, often lacks the capacity to handle these events effectively. In contrast, green infrastructure provides a more adaptive approach by slowing down and absorbing excess water. Restored wetlands and urban forests, for example, act as natural buffers, storing floodwaters and reducing the impact on built environments. By incorporating green infrastructure into flood management strategies, cities can better withstand and recover from extreme weather events.

Enhancing Water Quality

Urban runoff carries pollutants such as oil, chemicals, and debris into waterways, degrading water quality and harming aquatic ecosystems. This challenge is further compounded by aging sewer systems that may discharge untreated wastewater during heavy rain events. Green infrastructure addresses these issues by filtering pollutants through vegetation and soil systems. For instance, bioswales and constructed wetlands trap sediments, absorb nutrients, and break down contaminants before they reach natural water bodies. Improving water quality not only supports healthier ecosystems but also reduces the burden on energy-intensive water treatment facilities.

Promoting Climate Adaptation

Climate change presents complex challenges for urban water systems, including prolonged droughts, more intense rainfall, and rising sea levels. Green infrastructure plays a vital role in helping cities adapt to these changes. During droughts, systems like green roofs and rainwater harvesting reduce demand on municipal water supplies by capturing and storing water for later use. In areas vulnerable to sea level rise, coastal green infrastructure, such as mangroves and vegetated dunes, provides natural protection against storm surges and erosion. By integrating these solutions into urban planning, cities can build adaptive capacity and reduce vulnerability to climate impacts.

Building Social and Economic Resilience

Water challenges often exacerbate social inequities, as vulnerable populations are disproportionately affected by flooding, water scarcity, and poor water quality. Green infrastructure projects can address these inequities by providing underserved communities with access to cleaner environments, improved water systems, and recreational green spaces. Additionally, the economic benefits of green infrastructure—such as reduced flood damage, lower water treatment costs, and increased property values—contribute to broader urban resilience. These advantages create a positive feedback loop, where enhanced water management strengthens the overall economic stability of cities.

Integrating Multifunctional Solutions

Urban resilience requires systems that can address multiple challenges simultaneously. Green infrastructure's multifunctionality makes it uniquely suited for this purpose. A single project, such as a restored wetland, can manage stormwater, improve water quality, provide wildlife habitat, and offer recreational opportunities. This holistic approach not only maximizes resource efficiency but also strengthens the interconnected systems that support urban resilience.

Chapter 2: Understanding Environmental Impact Bonds

This chapter provides a comprehensive overview of EIBs and their role in financing green infrastructure. It begins by explaining the structure and mechanics of EIBs, highlighting how they differ from traditional financing methods. The discussion then explores the benefits of this model, including its ability to attract private investment, enhance accountability, and encourage outcome-driven projects. Finally, the chapter examines the challenges and limitations associated with EIBs, offering insights into how cities can effectively leverage this innovative tool. By understanding the principles and potential of EIBs, stakeholders can unlock new opportunities to fund sustainable urban water management and build resilient cities for the future.

Overview of EIBs and Their Role

EIBs are an innovative financial instrument designed to fund environmental projects by tying payments to measurable outcomes. Unlike traditional municipal bonds that provide fixed returns regardless of project success, EIBs incorporate performance-based metrics to determine returns for investors. This structure incentivizes the successful implementation of projects while sharing financial risks among stakeholders.

EIBs operate on a pay-for-success model, where investors provide upfront capital for a project, and repayments are contingent on the project achieving predefined environmental or social outcomes. For example, a city may issue an EIB to fund a green infrastructure project aimed at reducing stormwater runoff. If the project meets or exceeds performance targets, investors receive their principal plus a success payment. If targets are not met, investors may only receive partial repayment, reflecting the shared risk.

One of the key roles of EIBs is to bridge funding gaps for sustainability projects, particularly those involving green infrastructure. Traditional funding mechanisms, such as government grants or public budgets, are often constrained by competing priorities and limited resources. EIBs attract private capital to projects that might otherwise struggle to secure funding. By linking financial returns to measurable impact, EIBs appeal to socially conscious investors seeking to combine financial performance with positive environmental outcomes.

Another critical role of EIBs is fostering innovation in project design and implementation. The performance-based structure encourages project developers to adopt data-driven approaches and cutting-edge technologies to maximize success. For instance, predictive analytics and monitoring systems are often integrated into EIB-funded projects to track progress and ensure targets are met. This emphasis on innovation not only improves project outcomes but also helps establish best practices that can be replicated in future initiatives.

EIBs also enhance accountability and transparency in environmental projects. Since repayments are tied to independently verified outcomes, stakeholders—including governments, investors, and the public—can have greater confidence in the effectiveness of funded initiatives. This accountability reduces the risk of misallocated resources and ensures that funds are directed toward projects with tangible benefits. Furthermore, the collaborative nature of EIBs fosters partnerships between public and private entities, leveraging diverse expertise to address complex environmental challenges.

In addition to their practical benefits, EIBs play a strategic role in advancing sustainability goals at the local, national, and global levels. For cities, EIBs offer a means to align financial strategies with sustainability plans, enabling the implementation of green infrastructure projects that contribute to climate resilience. On a broader scale, EIBs support progress toward global objectives like the United Nations SDGs, particularly those related to clean water (SDG 6), sustainable cities (SDG 11), and climate action (SDG 13).

Despite their promise, EIBs are not without challenges. The complexity of structuring these bonds, the need for robust data to define and measure outcomes, and the relatively small pool of investors familiar with this model can limit their application. However, as awareness grows and more successful examples emerge, EIBs are expected to become an increasingly viable tool for financing environmental projects.

In summary, EIBs represent a forward-thinking approach to funding green infrastructure and other sustainability initiatives. By aligning financial returns with measurable outcomes, they attract private capital, promote innovation, and enhance accountability, making them a valuable instrument in the transition toward resilient and sustainable urban systems. Their role in bridging funding gaps and advancing global sustainability goals underscores their importance in modern environmental finance.

Structure and Stakeholder Roles

The structure of EIBs is designed to align financial resources with environmental outcomes while distributing risks and rewards among stakeholders. EIBs are structured as performance-based contracts, with returns tied to the success of a specific project. This innovative model requires the active involvement of several key stakeholders, each playing a critical role in ensuring the bond's success.

Issuer

The issuer of an EIB is typically a public entity, such as a municipal government, utility, or environmental agency. The issuer identifies the environmental problem to be addressed and designs a project to achieve specific outcomes. For example, a city may issue an EIB to fund green infrastructure initiatives, such as rain gardens or permeable pavements, aimed at reducing stormwater runoff and mitigating flooding. The issuer's responsibilities include defining project goals, developing performance metrics, and coordinating with other stakeholders to structure the bond.

Investors

Investors provide the upfront capital needed to fund the project. These are often private entities, such as institutional investors, impact investment funds, or philanthropic organizations. The investors' role is to take on financial risk in exchange for potential returns based on the project's performance. Unlike traditional bonds with fixed returns, EIBs offer variable payouts tied to outcomes. If the project meets or exceeds the agreed-upon performance metrics, investors may receive a premium return. If the project underperforms, their returns may be reduced, reflecting the shared risk.

Service Provider

The service provider is responsible for implementing and managing the project funded by the EIB. This can include engineering firms, environmental organizations, or local contractors with expertise in the relevant field. For instance, in an EIB aimed at stormwater management, the service provider might design and install green infrastructure systems such as bioswales or urban wetlands. The service provider works closely with the issuer to ensure that the project is completed on time and in line with the agreed-upon objectives.

Independent Evaluator

An independent evaluator plays a critical role in verifying the project's performance. This stakeholder is responsible for monitoring and assessing whether the project achieves the predefined outcomes, such as a reduction in stormwater runoff or improved water quality. The evaluator uses data-driven methods to measure results, ensuring transparency and accountability. Their findings determine whether the issuer makes payments to investors based on the project's success. The inclusion of an independent evaluator builds trust among stakeholders and ensures that outcomes are objectively assessed.

Beneficiaries

While not directly involved in the bond's financial structure, beneficiaries—such as local residents, businesses, and ecosystems—are central to the purpose of the EIB. Beneficiaries experience the environmental, social, and economic benefits of the funded project, such as reduced flooding, cleaner water, and enhanced urban green spaces. Their support is often critical to the project's acceptance and long-term sustainability.

Financial Intermediary

A financial intermediary, such as a bank or investment firm, often facilitates the issuance of the EIB. This entity provides expertise in structuring the bond, attracting investors, and managing financial transactions. The intermediary ensures that the bond is designed to meet the needs of both issuers and investors, balancing risk and return in a way that aligns with the project's goals.

EIB Structure in Practice

The EIB process typically begins with the issuer identifying an environmental challenge and designing a project with measurable outcomes. Investors then provide upfront funding, which is used by the service provider to implement the project. Throughout the project's duration, the independent evaluator monitors progress against the agreed-upon metrics. Based on the results, the issuer makes payments to investors, which may include success payments if the project exceeds expectations.

Risk-Sharing and Outcome Metrics in Environmental Impact Bonds

One of the defining features of EIBs is their ability to distribute financial risks and rewards among stakeholders based on project performance. This pay-for-success model ensures that all parties are incentivized to achieve positive outcomes, while the use of robust outcome metrics provides transparency and accountability. Understanding the mechanisms of risk-sharing and the importance of outcome metrics is essential for the effective design and implementation of EIBs.

Risk-Sharing Mechanisms

Risk-sharing in EIBs is structured to align the interests of stakeholders, including the issuer, investors, and service providers. By tying payments to measurable outcomes, EIBs ensure that financial rewards reflect the success of the project. Here's how the risk-sharing mechanism typically works:

• **Investors' Role in Risk-Taking**: Investors provide the upfront capital required to implement the project, taking on the initial financial risk. Their returns depend on the project's performance. If the project achieves or exceeds the predetermined targets, investors may receive a premium return. Conversely, if the project underperforms, they may only receive partial repayment, or in some cases, none at all. This model incentivizes investors to back projects with a high likelihood of success.

• **Issuer's Role in Risk Mitigation**: The issuer, often a municipal government or public agency, benefits from reduced financial exposure. Instead of paying the full cost of the project upfront, the issuer only makes payments if the project meets its objectives. This arrangement allows public entities to allocate resources efficiently while ensuring accountability for results.

• **Shared Responsibility Among Stakeholders**: Service providers are indirectly involved in the risk-sharing process. Their performance in designing and implementing the project directly affects whether outcome metrics are achieved. While they are typically paid through the initial investment, their reputation and future opportunities may depend on their ability to deliver successful results.

Benefits of Risk-Sharing

The risk-sharing model of EIBs offers several advantages:

• **Encourages Innovation**: By shifting some financial risk to investors, issuers can pursue innovative projects that may not have been feasible under traditional funding models.

• **Reduces Public Financial Burden**: Municipalities and public agencies can fund critical environmental projects without bearing the entire financial risk.

• **Aligns Incentives**: All stakeholders have a vested interest in the project's success, creating a collaborative environment focused on achieving measurable outcomes.

Outcome Metrics: The Foundation of Accountability

Outcome metrics are the measurable indicators used to assess the success of an EIB-funded project. These metrics serve as the foundation for determining payments to investors and ensuring that the project delivers tangible environmental or social benefits. The selection of appropriate outcome metrics is critical to the bond's success.

• **Defining Outcome Metrics**: Metrics must be specific, measurable, achievable, relevant, and time-bound (SMART). For example, a stormwater management project funded by an EIB might use metrics such as the percentage reduction in stormwater runoff volume, improvements in water quality, or the number of flood events mitigated.

• **Monitoring and Verification**: Independent evaluators play a key role in monitoring progress and verifying outcomes. They collect data, analyze performance, and ensure that metrics are met in accordance with the bond's terms. This independent verification builds trust among stakeholders and ensures transparency.

• **Adjusting Metrics for Context**: Outcome metrics should be tailored to the specific goals and conditions of each project. For example, a green infrastructure project in an urban area prone to flooding may prioritize metrics related to flood risk reduction, while a project in a water-scarce region may focus on metrics related to water conservation and groundwater recharge.

Challenges in Defining and Measuring Outcomes

Despite their importance, defining and measuring outcome metrics in EIBs can be complex. Common challenges include:

• **Data Availability**: Reliable baseline data is necessary to establish benchmarks and track progress. In some cases, such data may be limited or unavailable.

• **Attribution**: It can be difficult to isolate the impact of a single project, particularly in systems influenced by multiple factors.

• **Long-Term Impact**: Some benefits of green infrastructure, such as improved ecosystem health, may take years to materialize, making short-term metrics less reflective of true success.

Balancing Risk and Reward

Striking the right balance between risk and reward is critical to the success of EIBs. The terms of the bond must be structured to provide fair incentives for investors while ensuring that the issuer does not overpay for uncertain results. Scenario analysis and sensitivity testing can help stakeholders design performance-based contracts that align with project goals and financial realities.

Limitations and Future Potential of Environmental Impact Bonds

EIBs have emerged as an innovative tool for financing sustainable projects, particularly those involving green infrastructure. However, like any financial mechanism, EIBs are not without their limitations. Understanding these challenges is essential for improving their design and expanding their application. Simultaneously, the model holds significant potential for growth as awareness and expertise in sustainable financing evolve.

Limitations of EIBs

• Complex Structuring Process

The design and implementation of an EIB require a high level of technical expertise, making the process complex and resource-intensive. Establishing measurable outcome metrics, negotiating contracts, and aligning stakeholder incentives involve extensive time, cost, and collaboration. Smaller municipalities or organizations with limited technical capacity may find it difficult to navigate these requirements.

• High Transaction Costs

The administrative and legal expenses associated with structuring an EIB can be significant, potentially outweighing the benefits for smaller-scale projects. Hiring financial intermediaries, independent evaluators, and legal advisors adds to the overall cost, making EIBs more suitable for large-scale initiatives with significant financial stakes.

• Limited Investor Base

EIBs currently appeal to a niche group of socially conscious investors and philanthropic organizations. The relatively small pool

of investors familiar with pay-for-success models limits the
scalability of EIBs. Furthermore, potential investors may be deterred
by the financial risks involved, particularly for projects with
uncertain or long-term outcomes.

• Challenges in Measuring Outcomes

Accurately defining and measuring outcomes can be a significant
challenge, especially for projects with complex or long-term
benefits. Factors such as data availability, baseline conditions, and
external influences can complicate the assessment of success.
Additionally, outcome metrics may not capture all co-benefits of
green infrastructure, such as community well-being or biodiversity
enhancement, limiting the perceived value of the project.

• Risk of Underperformance

EIBs transfer some financial risk to investors, but underperformance
can have broader implications. If projects fail to meet their targets,
public entities may face reputational damage, and investors may lose
confidence in the model. This risk can discourage both issuers and
investors from pursuing EIBs in the future.

Future Potential of EIBs

Despite these limitations, EIBs hold significant potential to expand
and transform the way cities and organizations fund sustainable
projects. With increased awareness, capacity-building, and
innovation, EIBs could become a mainstream tool for environmental
financing.

• Scaling to New Markets

As more cities and organizations recognize the benefits of EIBs, the
model can be adapted to diverse contexts, including developing
countries and smaller municipalities. Expanding access to technical

expertise and simplifying the structuring process will help make EIBs more accessible to a broader range of stakeholders.

• Integration with Other Financial Tools

EIBs can be combined with other financing mechanisms, such as green bonds, public-private partnerships, and grants, to create hybrid models that maximize impact. For instance, pairing EIBs with grants could help lower upfront costs for public entities, reducing barriers to adoption.

• Broadening the Investor Base

Efforts to educate and engage a wider range of investors can help grow the market for EIBs. Institutional investors, such as pension funds and insurance companies, could be drawn to EIBs by highlighting their potential for stable returns and alignment with Environmental, Social, and Governance (ESG) criteria. Developing standardized frameworks and tools for EIBs would also enhance investor confidence.

• Advancing Technology and Data Analytics

Technological advancements can improve the efficiency and effectiveness of EIBs. For example, predictive modeling, remote sensing, and real-time monitoring tools can provide more accurate and reliable data for measuring outcomes. Blockchain technology could enhance transparency and accountability by securely recording performance metrics and financial transactions.

• Expanding Outcome Metrics

Broadening the scope of outcome metrics to include co-benefits, such as social equity, public health improvements, and biodiversity gains, can increase the perceived value of EIB-funded projects. This approach could also attract a wider range of stakeholders who value these additional benefits.

• Building Institutional Capacity

Capacity-building initiatives, such as training programs and knowledge-sharing platforms, can help public agencies and organizations better understand and implement EIBs. Partnerships with academic institutions, think tanks, and industry experts can play a crucial role in advancing expertise in this area.

Chapter 3: Public Funding Mechanisms for Green Infrastructure

This chapter explores the critical role of public funding in advancing green infrastructure projects. It begins by examining the various mechanisms that governments use to allocate funds, such as grants, subsidies, and dedicated budgets. The discussion highlights how these tools can be used to catalyze green infrastructure development and incentivize stakeholder participation. Additionally, the chapter addresses the challenges associated with public funding, including competing priorities, limited budgets, and regulatory hurdles, and offers insights into strategies for overcoming these barriers.

Role of Governments in Green Financing

Governments play a critical role in financing green infrastructure, serving as both enablers and direct funders of sustainable urban water management projects. Their involvement ensures that essential resources are allocated effectively, addressing the challenges of climate change, urbanization, and water scarcity while promoting environmental, social, and economic resilience.

Funding Green Infrastructure Projects

Governments at the federal, state, and municipal levels are primary providers of funding for green infrastructure initiatives. Through grants, subsidies, and direct investment, they allocate resources to projects that mitigate stormwater runoff, reduce flooding, and improve water quality. For instance, municipalities often establish dedicated budgets for green infrastructure as part of broader urban sustainability plans, ensuring that funding is consistently directed toward long-term goals.

Public funding is particularly crucial in areas where private investment is limited or where projects serve public interests, such as reducing flood risks or enhancing community spaces. By providing the initial capital for green infrastructure, governments can help overcome financial barriers and incentivize innovation in sustainable urban development.

Policy and Regulatory Frameworks

Governments create the policy and regulatory environments that enable green infrastructure financing. Policies such as stormwater management regulations, environmental impact assessments, and incentives for green building certifications encourage the adoption of sustainable practices. For example, many cities implement stormwater utility fees, which generate revenue specifically for green infrastructure projects. These fees create a steady funding stream while also encouraging property owners to adopt practices that reduce runoff, such as installing permeable pavements or rain gardens.

In addition, governments can introduce tax incentives, credits, or deductions for developers and businesses that invest in green infrastructure. These measures not only reduce the financial burden on private stakeholders but also align public and private interests toward achieving sustainability objectives.

Facilitating Public-Private Partnerships

Governments are instrumental in fostering PPPs to finance green infrastructure projects. By collaborating with private sector entities, governments can leverage additional resources and expertise while sharing risks and responsibilities. For example, a city may partner with private developers to integrate green roofs or bioswales into urban redevelopment projects. In such arrangements, governments often provide financial support, such as matching funds or loan guarantees, to attract private investment and ensure project viability.

These partnerships allow for cost-sharing, ensuring that public funds are used efficiently while enabling large-scale implementation of green infrastructure solutions. Governments also play a coordinating role, ensuring that stakeholders work together effectively to meet project goals.

Supporting Innovation and Research

Governments contribute to the advancement of green infrastructure by funding research and pilot projects that demonstrate the viability of innovative solutions. By investing in feasibility studies, environmental assessments, and monitoring programs, they provide the data needed to inform decision-making and refine best practices. Research grants and funding for academic partnerships further support the development of new technologies and methods that enhance the performance and cost-effectiveness of green infrastructure.

In addition, governments often serve as early adopters of green technologies, implementing demonstration projects in public spaces such as parks, schools, and government buildings. These initiatives showcase the benefits of green infrastructure to the broader community, encouraging wider adoption by private stakeholders.

Promoting Equity in Green Infrastructure

Governments play a vital role in ensuring that green infrastructure benefits are equitably distributed across communities. By prioritizing investments in underserved or vulnerable neighborhoods, they address disparities in access to clean water, flood protection, and green spaces. For example, targeted funding programs can support the installation of green infrastructure in low-income areas, reducing the risk of flooding and improving environmental quality for residents.

Additionally, governments can engage local communities in the planning and implementation of green infrastructure projects, fostering public support and ensuring that solutions are tailored to community needs.

Key Funding Programs for Green Infrastructure (Grants, Subsidies, etc.)

Governments and public agencies offer a variety of funding programs to support the development and implementation of green infrastructure projects. These programs provide critical financial resources to address the challenges of urban water management, such as stormwater runoff, flood mitigation, and water quality improvement. By leveraging tools like grants, subsidies, and tax incentives, public entities can catalyze green infrastructure investments and encourage stakeholder participation.

Grants for Green Infrastructure Projects

Grants are one of the most common funding mechanisms used to support green infrastructure initiatives. These funds are typically allocated by federal, state, or local governments and do not require repayment. Grants are often targeted at specific goals, such as stormwater management, urban greening, or climate adaptation, and are designed to lower financial barriers for project implementation.

• **Federal Grants**: Many national governments provide large-scale grants for environmental projects. For example, in the United States, the Environmental Protection Agency (EPA) offers funding through programs like the Clean Water State Revolving Fund (CWSRF) and the Water Infrastructure Finance and Innovation Act (WIFIA). These programs support municipalities in implementing green infrastructure solutions to improve water quality and manage stormwater.

• **Local and Regional Grants**: Municipal governments and regional agencies also allocate grants for green infrastructure projects. These grants often focus on community-specific needs, such as flood prevention or urban heat island reduction. Local programs may prioritize smaller projects, such as rain gardens or permeable pavement installations, to encourage widespread adoption of green practices.

• **Competitive Grants**: Some funding opportunities are awarded through competitive processes, requiring applicants to demonstrate the potential impact and feasibility of their projects. This approach encourages innovation and ensures that funds are directed toward high-quality initiatives.

Subsidies and Incentive Programs

Subsidies provide financial assistance to offset the costs of green infrastructure projects, making them more accessible for property owners, developers, and businesses. Incentive programs, on the other hand, encourage stakeholders to adopt sustainable practices by offering financial rewards or cost reductions.

• **Stormwater Fee Credits**: Many cities implement stormwater fee credit programs, allowing property owners to reduce their utility fees by installing green infrastructure solutions, such as rain barrels, green roofs, or bioswales. These programs incentivize private investment in sustainable practices while generating public benefits through reduced stormwater runoff.

• **Rebates for Green Infrastructure**: Rebates are commonly offered for the installation of green infrastructure features. For example, homeowners may receive rebates for planting trees, replacing impervious surfaces with permeable materials, or implementing rainwater harvesting systems. These subsidies encourage small-scale projects that collectively contribute to larger environmental goals.

• **Tax Incentives**: Governments can use tax incentives to promote green infrastructure development. For instance, property tax credits or deductions may be offered to businesses or developers that incorporate green roofs, constructed wetlands, or other sustainable features into their projects. These incentives lower the financial burden of adopting green practices and encourage widespread participation.

Publicly-Funded Loan Programs

In addition to grants and subsidies, governments offer low-interest loan programs to finance green infrastructure projects. These loans provide upfront capital for large-scale initiatives and allow borrowers to repay the funds over time, often with favorable terms.

• **Revolving Loan Funds**: Programs like the Clean Water State Revolving Fund (CWSRF) provide loans to municipalities for wastewater and stormwater infrastructure projects. The revolving nature of these funds ensures that resources are replenished as loans are repaid, creating a sustainable funding cycle.

• **Green Bonds**: Some public agencies issue green bonds to raise capital for environmentally sustainable projects. These bonds attract socially conscious investors and allow governments to finance green infrastructure initiatives without depleting public budgets.

Targeted Funding for Underserved Communities

Many funding programs are designed to address equity concerns by prioritizing investments in underserved or vulnerable communities. These programs aim to reduce environmental disparities and ensure that all residents benefit from green infrastructure projects.

• **Environmental Justice Grants**: Governments and organizations provide funding specifically for projects in low-income or historically marginalized communities. These grants support initiatives that improve water quality, reduce flood risks, and create green spaces in areas that have traditionally lacked investment.

• **Community-Based Programs**: Public agencies often collaborate with local organizations to implement green infrastructure projects at the neighborhood level. These partnerships empower communities to participate in decision-making and ensure that projects align with local needs.

Research and Pilot Project Funding

Governments also fund research and pilot projects to demonstrate the feasibility and benefits of green infrastructure. These programs provide financial support for studying innovative solutions, testing new technologies, and developing best practices. Successful pilot projects often serve as models for broader implementation and attract additional funding from public and private sources.

Challenges in Public Funding for Green Infrastructure

Public funding is a critical component of financing green infrastructure projects, enabling cities to address urban water management challenges while promoting sustainability. However, relying on public resources comes with several challenges that can limit the effectiveness and scalability of green infrastructure initiatives. These challenges often stem from financial constraints, competing priorities, administrative complexities, and public perception.

Limited Budgets and Competing Priorities

One of the primary challenges in public funding is the limited availability of financial resources. Municipal and state budgets are often constrained, with funds allocated to multiple essential services, such as healthcare, education, and transportation. As a result, funding for green infrastructure must compete with other priorities, making it difficult to secure dedicated resources.

Additionally, green infrastructure projects often require significant upfront investment, even though their long-term benefits, such as flood mitigation and improved water quality, may take years to materialize. This disconnect between short-term costs and long-term gains can make it harder for public officials to justify allocating funds to green infrastructure, particularly in resource-limited communities.

Inconsistent Funding Streams

Green infrastructure projects rely on stable and predictable funding to ensure successful implementation and maintenance. However, public funding sources are often inconsistent, influenced by political cycles, economic fluctuations, and policy changes. For example, a change in government priorities or a reduction in tax revenue during economic downturns can disrupt funding commitments, delaying or halting projects.

This uncertainty can also deter private sector participation, as inconsistent public support may signal higher risks for collaborative projects. Long-term funding mechanisms, such as stormwater utility fees or dedicated environmental funds, can help address this challenge, but they require political will and community buy-in to implement effectively.

Administrative and Regulatory Barriers

Public funding for green infrastructure is often subject to complex administrative and regulatory processes. Securing grants, subsidies, or loans involves navigating extensive application procedures, meeting eligibility criteria, and adhering to reporting requirements. These processes can be time-consuming and resource-intensive, particularly for smaller municipalities or organizations with limited administrative capacity.

In some cases, regulatory frameworks are better suited to traditional gray infrastructure, creating additional hurdles for green infrastructure projects. For example, outdated building codes or stormwater regulations may not fully accommodate nature-based solutions, requiring extra effort to secure permits or demonstrate compliance.

Maintenance and Lifecycle Costs

While public funding often focuses on the initial implementation of green infrastructure, ongoing maintenance and lifecycle costs are frequently overlooked. Projects like green roofs, bioswales, and urban wetlands require regular upkeep to ensure they function effectively over time. Without adequate funding for maintenance, the performance and benefits of green infrastructure can degrade, undermining its long-term value.

Allocating resources for maintenance can be challenging, as these costs are less visible and less politically appealing than new project announcements. Developing comprehensive funding plans that account for lifecycle costs is essential to address this issue, but it requires proactive planning and sustained commitment from public agencies.

Public Perception and Support

Gaining public support for green infrastructure funding can be challenging, particularly when residents are unfamiliar with the concept or skeptical of its benefits. Some communities may perceive green infrastructure projects as unnecessary or overly expensive compared to traditional approaches. This lack of understanding can lead to resistance, particularly if funding mechanisms, such as stormwater fees or tax increases, directly impact residents.

Educating the public about the environmental, economic, and social benefits of green infrastructure is critical to building support. Transparent communication, community engagement, and demonstration projects can help address misconceptions and foster trust in public funding initiatives.

Equity and Accessibility Issues

Ensuring that public funding for green infrastructure is distributed equitably is another challenge. Historically underserved or marginalized communities often face greater environmental risks, such as flooding and water contamination, but may struggle to access funding for green infrastructure projects. Addressing this inequity requires targeted funding programs and policies that prioritize investments in vulnerable areas.

Strategies for Effective Public Investment in Green Infrastructure

Effective public investment in green infrastructure is essential for addressing urban water management challenges while promoting sustainability and resilience. By adopting strategic approaches, governments can maximize the impact of their funding, overcome financial and administrative barriers, and ensure equitable access to green infrastructure benefits.

Establishing Dedicated Funding Mechanisms

One of the most effective strategies for ensuring consistent public investment is the establishment of dedicated funding mechanisms for green infrastructure. Examples include stormwater utility fees, environmental impact fees, and dedicated tax revenue streams. These mechanisms provide a stable and predictable source of funding that is earmarked for specific purposes, reducing reliance on general municipal budgets.

Stormwater utility fees, for instance, charge property owners based on the impervious surfaces on their properties, which contribute to runoff. This approach not only generates revenue but also incentivizes property owners to adopt green infrastructure solutions, such as permeable pavements or rain gardens, to reduce fees.

Leveraging Public-Private Partnerships

Public-private partnerships (PPPs) enable governments to share the financial burden of green infrastructure projects with private sector entities. By collaborating with developers, businesses, and investors, public agencies can access additional resources and expertise while reducing upfront costs.

PPPs are particularly effective in large-scale projects where private entities can contribute funding or in-kind services in exchange for incentives such as tax credits, recognition, or shared benefits. For example, municipalities may partner with real estate developers to integrate green roofs, bioswales, and tree canopies into urban redevelopment projects.

Prioritizing Cost-Effective Solutions

Maximizing the impact of public investment requires careful prioritization of cost-effective green infrastructure solutions. Governments should assess the potential benefits and lifecycle costs of different projects, focusing on initiatives that deliver the greatest environmental, social, and economic returns. Tools such as cost-benefit analysis and lifecycle assessments can help identify high-impact projects.

For instance, small-scale, distributed solutions like rain barrels and tree plantings may offer significant cumulative benefits at a relatively low cost. Similarly, hybrid approaches that combine green and gray infrastructure can optimize performance while managing expenses.

Integrating Green Infrastructure into Urban Planning

Embedding green infrastructure into broader urban planning processes ensures that public investment aligns with long-term sustainability and resilience goals. Governments can integrate green infrastructure requirements into zoning codes, building regulations, and land-use plans to promote widespread adoption.

For example, cities can mandate the inclusion of green infrastructure elements, such as permeable pavements or green roofs, in new developments and retrofits. Additionally, incorporating green infrastructure into transportation projects, such as road medians and sidewalks, ensures that investments serve multiple purposes, including stormwater management and urban beautification.

Fostering Community Engagement

Engaging local communities in the planning, implementation, and maintenance of green infrastructure projects can enhance public investment effectiveness. Community involvement ensures that projects address local needs and priorities, increasing their acceptance and long-term success.

Governments can organize workshops, public consultations, and educational campaigns to inform residents about the benefits of green infrastructure and gather input on proposed projects. Offering financial incentives, such as grants or rebates for residential green infrastructure installations, also encourages community participation and amplifies the impact of public investment.

Addressing Equity and Inclusivity

Ensuring equitable distribution of public investment in green infrastructure is critical to addressing environmental justice concerns. Governments should prioritize funding for underserved communities that face disproportionate risks from flooding, water pollution, and lack of green spaces.

Targeted funding programs, such as grants for low-income neighborhoods or investments in community-driven projects, can address these disparities. Additionally, ensuring that green infrastructure benefits, such as improved air quality and reduced heat, reach vulnerable populations strengthens social resilience.

Enhancing Funding Synergies

To stretch public resources further, governments can align green infrastructure funding with other policy priorities, such as climate adaptation, public health, and economic development. For example, funding for climate resilience projects can include green infrastructure components that address stormwater management and reduce urban heat islands.

Similarly, integrating green infrastructure into public health initiatives, such as creating parks and recreational spaces, ensures that investments deliver multiple co-benefits. Leveraging co-funding opportunities across sectors maximizes the efficiency of public investment.

Chapter 4: Private Sector Participation in Green Infrastructure

This chapter explores the diverse ways in which the private sector contributes to green infrastructure development. It begins by examining the incentives that motivate businesses to invest in sustainable practices and the role of PPPs in facilitating collaboration. The discussion then highlights the challenges of private sector involvement, including regulatory barriers and financing gaps, while offering strategies to overcome these hurdles.

Rise of Public-Private Partnerships in Green Infrastructure

PPPs have emerged as a powerful mechanism for financing and implementing green infrastructure projects. By leveraging the resources, expertise, and innovation of both public and private sectors, PPPs address the challenges of urban water management while promoting sustainable development. This collaborative model not only bridges funding gaps but also accelerates the adoption of green solutions such as rain gardens, permeable pavements, and urban forests.

What Are PPPs?

PPPs are contractual agreements between public entities, such as municipalities or government agencies, and private organizations, including corporations, developers, or financial institutions. These partnerships allocate responsibilities, risks, and rewards between the parties, ensuring that each contributes to the project's success. In the context of green infrastructure, PPPs often involve private sector participation in funding, designing, constructing, operating, or maintaining sustainable urban water management systems.

PPPs differ from traditional procurement methods by emphasizing shared accountability and innovation. While the public sector provides oversight, regulation, and strategic goals, the private sector brings technical expertise, financial capital, and operational efficiency. This collaboration creates a balanced approach that aligns public and private interests.

Why Are PPPs Rising in Popularity?

Several factors have contributed to the growing prominence of PPPs in green infrastructure:

• **Funding Challenges**: Municipal budgets are often constrained, limiting the ability to finance large-scale green infrastructure projects independently. PPPs enable public entities to access private capital, reducing the financial burden on taxpayers while ensuring project feasibility.

• **Demand for Innovation**: Private companies frequently adopt cutting-edge technologies and approaches to maximize efficiency and outcomes. By partnering with the private sector, governments can integrate innovative solutions into green infrastructure projects.

• **Shared Risk**: PPPs distribute financial and operational risks between public and private partners. For instance, the private sector may assume responsibility for construction costs or performance guarantees, reducing the public sector's exposure to potential setbacks.

• **Scalability**: The collaborative nature of PPPs facilitates large-scale implementation of green infrastructure, which might otherwise be challenging for public entities to achieve on their own. This scalability is particularly important in addressing complex urban challenges, such as stormwater management and flood mitigation.

How PPPs Work in Green Infrastructure

In a typical green infrastructure PPP, the public sector defines project goals and establishes a regulatory framework, while the private sector takes on responsibilities such as financing, construction, and maintenance. Common structures include:

• **Design-Build-Operate-Maintain (DBOM)**: The private sector designs, builds, operates, and maintains the green infrastructure project for a specified period. The public sector retains ownership but benefits from the private sector's expertise.

• **Concession Agreements**: The public sector grants the private entity the right to operate and maintain a green infrastructure asset for a set duration. In return, the private partner may collect user fees or receive performance-based payments.

• **Co-Funding Models**: Both public and private entities contribute financial resources to the project, sharing costs and risks. This model is often used for community-based initiatives, such as urban parks or green roofs.

Examples of Green Infrastructure PPPs

PPPs have been successfully employed in various green infrastructure projects, demonstrating their versatility and effectiveness:

• **Stormwater Management**: In many cities, private developers work with municipalities to integrate green infrastructure into new developments. For example, developers may install permeable pavements and bioswales to comply with stormwater regulations, with public agencies providing technical support or financial incentives.

• **Green Roof Initiatives**: Governments often partner with private property owners to promote green roofs. Through financial incentives such as grants or tax credits, the public sector encourages private investment in urban greening, while the private sector benefits from reduced energy costs and enhanced property value.

• **Urban Parks and Green Spaces**: PPPs have facilitated the creation of public green spaces by combining public funding with private sponsorship or philanthropic contributions. These projects enhance urban resilience while offering recreational and aesthetic benefits.

Challenges and Opportunities

While PPPs offer significant advantages, they also face challenges:

• **Regulatory Barriers**: Complex permitting processes or unclear regulations can hinder the formation of PPPs. Governments need to establish clear guidelines to encourage collaboration.

• **Stakeholder Alignment**: Public and private entities may have different priorities. Effective communication and contract design are essential to align goals and ensure mutual benefits.

• **Long-Term Accountability**: Ensuring that private partners uphold their responsibilities over the project's lifecycle requires robust monitoring and performance metrics.

Despite these challenges, PPPs hold immense potential to drive green infrastructure development. With the right policies, frameworks, and incentives, they can serve as a cornerstone of sustainable urban water management.

Incentives for Private Investment in Green Infrastructure

Encouraging private sector investment in green infrastructure is critical for addressing the growing challenges of urban water management and climate resilience. Governments and public agencies have developed various incentives to attract private investment, ensuring that businesses, developers, and investors see tangible benefits in contributing to sustainable solutions. These incentives range from financial and regulatory benefits to reputational gains, collectively fostering a favorable environment for private sector participation.

Financial Incentives

Financial incentives are among the most effective tools for encouraging private investment in green infrastructure. They reduce the financial burden on businesses and developers, making sustainable projects more economically viable.

• **Grants and Subsidies**: Governments often provide direct funding to private entities through grants and subsidies. For example, grants may cover a portion of the costs associated with installing green roofs, permeable pavements, or rainwater harvesting systems. These financial supports lower upfront expenses and encourage businesses to adopt green practices.

• **Tax Incentives**: Tax-based incentives, such as credits, deductions, or exemptions, make green infrastructure investments more attractive. For instance, property owners who install sustainable features like bioswales or vegetative barriers may qualify for property tax reductions. Similarly, businesses that integrate green infrastructure into their facilities may benefit from income tax credits, offsetting project costs.

• **Low-Interest Loans and Revolving Funds**: Public agencies often offer low-interest loans or access to revolving funds to finance green infrastructure projects. These programs provide affordable capital to private investors, enabling them to implement large-scale sustainable solutions without significant financial strain.

Regulatory Incentives

Regulatory frameworks can be tailored to encourage private sector investment by offering flexibility and benefits for compliance with sustainability goals.

• **Stormwater Fee Credits**: Many cities charge stormwater management fees based on the impervious surface area of properties. To incentivize green infrastructure, property owners may receive credits or reduced fees for implementing sustainable practices that reduce runoff, such as installing rain gardens or permeable pavements. These credits not only lower costs for businesses but also align private interests with public sustainability objectives.

• **Expedited Permitting**: Regulatory processes can be streamlined for projects incorporating green infrastructure. Developers who include sustainable features in their designs may qualify for fast-tracked permitting, reducing delays and administrative burdens. This incentive is particularly valuable in urban areas where lengthy approval processes can impact project timelines and profitability.

• **Zoning and Density Bonuses**: Governments can offer zoning incentives to encourage green infrastructure integration. For instance, developers may be allowed increased building heights or higher floor-area ratios in exchange for incorporating features like green roofs, urban wetlands, or stormwater retention systems. These bonuses enhance the economic return on investments while promoting sustainability.

Reputational Benefits

Private entities increasingly recognize the value of aligning with sustainability goals to enhance their brand image and attract socially conscious consumers and investors.

• **Corporate Social Responsibility (CSR)**: Investing in green infrastructure aligns with corporate social responsibility objectives, showcasing a company's commitment to environmental stewardship. Businesses that adopt sustainable practices often gain a competitive edge by appealing to environmentally conscious customers and investors.

• **Green Certifications and Ratings**: Projects incorporating green infrastructure may qualify for certifications such as LEED (Leadership in Energy and Environmental Design) or SITES (Sustainable Sites Initiative). These certifications enhance a company's reputation and marketability by demonstrating adherence to high sustainability standards.

• **Recognition Programs**: Public agencies and industry groups often establish recognition programs to highlight businesses that invest in green infrastructure. Awards, public acknowledgments, and inclusion in sustainability reports can boost a company's profile and encourage other private entities to follow suit.

Market Opportunities

Private investment in green infrastructure can unlock new revenue streams and market opportunities, making sustainability an economically attractive option.

• **Energy Savings**: Features like green roofs and urban tree canopies reduce building energy consumption by improving insulation and shading. These energy savings translate to reduced operating costs, offering a direct financial return on investment.

• **Increased Property Value**: Properties with integrated green infrastructure often command higher market values due to their enhanced aesthetics, environmental benefits, and compliance with sustainability trends. Developers and property owners can capitalize on this premium to justify their investment.

• **New Business Models**: Green infrastructure opens avenues for innovative business models, such as water reuse systems or stormwater credit trading markets. Businesses can monetize sustainable practices by participating in these emerging markets, creating additional financial incentives.

Collaborative Funding and Public-Private Partnerships

Collaboration between public and private entities can amplify the impact of green infrastructure investments while reducing risks for businesses.

• **Cost-Sharing Models**: PPPs often involve cost-sharing arrangements, where governments provide partial funding or technical assistance for green infrastructure projects. This reduces the financial burden on private investors while ensuring that public sustainability goals are met.

• **Performance-Based Contracts**: In some cases, private entities receive payments based on the success of their green infrastructure initiatives, such as stormwater reduction or improved water quality. These performance-based models incentivize businesses to optimize the effectiveness of their investments.

Challenges and Considerations

Despite the array of incentives, private sector engagement in green infrastructure can face challenges, including high upfront costs, long payback periods, and limited awareness of available benefits. Addressing these barriers requires targeted outreach, education, and collaboration between public and private stakeholders.

Integrating Green Infrastructure in Private Projects

The integration of green infrastructure into private development projects is becoming increasingly common as businesses, developers, and property owners recognize the environmental, social, and economic benefits of sustainable design. By incorporating features such as green roofs, rain gardens, permeable pavements, and bioswales, private entities contribute to urban resilience while enhancing the value and functionality of their properties. This integration reflects a growing alignment between private sector goals and sustainability priorities.

Drivers for Integration

Private entities are motivated to include green infrastructure in their projects for several reasons:

• **Regulatory Compliance**: Many municipalities require private developers to meet stormwater management standards or environmental impact regulations. Green infrastructure provides an effective way to comply with these requirements, offering solutions that mitigate runoff, improve water quality, and reduce the risk of flooding.

• **Cost Savings**: Green infrastructure often leads to long-term savings by reducing stormwater utility fees, energy costs, and the need for gray infrastructure upgrades. For example, green roofs improve insulation, lowering heating and cooling expenses, while permeable pavements reduce wear and tear on drainage systems.

• **Market Demand**: Consumers and tenants increasingly prioritize sustainability in their purchasing and leasing decisions. Properties featuring green infrastructure are more attractive to environmentally conscious buyers and renters, enhancing marketability and property value.

• **CSR**: Businesses seeking to demonstrate their commitment to sustainability often incorporate green infrastructure into their facilities as part of broader CSR initiatives. This alignment with environmental goals improves brand reputation and fosters customer loyalty.

Common Applications of Green Infrastructure

Green infrastructure can be seamlessly integrated into a variety of private development projects, including commercial, residential, and industrial properties.

• **Green Roofs and Walls**: These features are particularly popular in urban areas, where they reduce building energy use, improve air quality, and provide aesthetic appeal. Green roofs also extend the lifespan of roofing materials by protecting them from UV exposure and temperature fluctuations.

• **Permeable Pavements**: Used in parking lots, driveways, and walkways, permeable pavements allow stormwater to infiltrate the ground, reducing runoff and recharging groundwater supplies. They are especially beneficial in areas with strict stormwater management regulations.

• **Rain Gardens and Bioswales**: These landscaped features capture and filter stormwater, preventing pollutants from entering waterways. They are commonly installed in residential developments, office complexes, and retail spaces to manage localized flooding and improve site aesthetics.

• **Retention Ponds and Wetlands**: Larger developments, such as industrial parks or shopping centers, often include retention ponds or constructed wetlands to manage stormwater on-site. These features provide habitat for wildlife while reducing the burden on municipal drainage systems.

Collaboration with Public Entities

Successful integration of green infrastructure often involves collaboration between private developers and public agencies. Municipalities can support private sector efforts through incentives, technical guidance, and streamlined permitting processes.

• **Incentives and Rebates**: Governments may offer financial incentives, such as tax credits or grants, to encourage the inclusion of green infrastructure in private projects. For example, a city might provide rebates for the installation of rainwater harvesting systems or tree plantings.

• **Technical Assistance**: Public agencies can provide expertise in green infrastructure design, helping developers navigate technical challenges and ensure compliance with local regulations. Collaboration with environmental consultants or urban planners can further streamline the process.

• **PPPs**: In some cases, private entities work with public agencies on joint green infrastructure projects, such as shared stormwater management systems or urban greening initiatives. These partnerships enable cost-sharing and broader community benefits.

Overcoming Barriers to Integration

While the benefits of green infrastructure are clear, private entities may encounter challenges when integrating these features into their projects. Common barriers include high upfront costs, limited knowledge of green infrastructure benefits, and uncertainty about long-term maintenance requirements.

• **Education and Awareness**: Providing developers and property owners with information about the economic and environmental advantages of green infrastructure can address misconceptions and increase adoption rates.

• **Financial Support**: Public funding programs, such as low-interest loans or cost-sharing arrangements, can help offset initial expenses and make green infrastructure more accessible for private projects.

• **Clear Guidelines**: Municipalities can facilitate integration by offering clear and consistent regulations, design standards, and maintenance requirements for green infrastructure.

Future Opportunities for Private Sector Collaboration in Green Infrastructure

As the demand for sustainable urban water management grows, opportunities for private sector collaboration in green infrastructure are expanding. These collaborations can drive innovation, attract investment, and foster scalable solutions that address pressing environmental challenges. By leveraging private sector expertise, resources, and networks, cities can accelerate the adoption of green infrastructure while creating shared benefits for businesses, communities, and the environment.

Advancing Public-Private Partnerships

PPPs will continue to play a pivotal role in green infrastructure development. Future opportunities lie in structuring innovative PPP models that distribute risks and rewards fairly. For instance, private entities can co-invest in projects such as green roofs, urban wetlands, or stormwater retention systems, with performance-based contracts ensuring mutual accountability. As cities refine their regulatory frameworks, PPPs can facilitate more integrated and large-scale green infrastructure initiatives.

Developing Green Financing Models

The private sector has an opportunity to collaborate on new financing mechanisms tailored to green infrastructure. These include green bonds, sustainability-linked loans, and impact investing platforms. Such models align financial returns with environmental outcomes, attracting a broader pool of investors, including institutional and socially conscious investors. By collaborating with governments and financial institutions, private entities can help design and implement innovative funding strategies.

Promoting Technology-Driven Solutions

The private sector can contribute to green infrastructure by advancing technology-driven solutions. Opportunities include developing smart systems for stormwater monitoring, using predictive analytics for flood risk assessment, and integrating Internet of Things (IoT) devices to optimize green infrastructure performance. Collaborative research and development (R&D) efforts between private firms, academic institutions, and public agencies can further enhance these technologies.

Scaling Community-Based Projects

Private businesses can partner with local organizations and municipalities to scale community-based green infrastructure projects. For example, companies can support urban tree-planting initiatives, sponsor green space development, or provide technical expertise for neighborhood rain garden programs. These collaborations foster local engagement while contributing to CSR objectives.

Chapter 5: Innovative Financing Models

This chapter examines innovative financing models for sustainable urban water management. It highlights green and climate resilience bonds, which attract capital for sustainability initiatives, and impact investing, which supports projects delivering financial and environmental returns. Crowdfunding and community financing are discussed as ways to engage local stakeholders and foster grassroots support. Finally, the chapter explores blending traditional and innovative approaches to overcome financial barriers and advance green infrastructure development at scale.

Green Bonds and Climate Resilience Bonds

Green bonds and climate resilience bonds are innovative financial instruments designed to fund projects that deliver measurable environmental benefits. By providing targeted capital for sustainability initiatives, these bonds are becoming vital tools for governments, corporations, and financial institutions aiming to address the challenges of urban water management and climate change. Their structured approach offers a compelling mechanism for mobilizing private and institutional investment in green infrastructure projects, such as stormwater management systems, permeable pavements, and urban wetlands.

Green Bonds: Definition and Purpose

Green bonds are debt securities issued to raise funds for projects with clear environmental objectives. Unlike traditional bonds, the proceeds from green bonds are earmarked exclusively for initiatives that contribute to sustainability goals. Examples include renewable energy, energy efficiency, and, in the context of green infrastructure, water management projects like rain gardens and bioswales. The key appeal of green bonds lies in their dual value proposition: they offer investors financial returns while enabling measurable environmental impact.

The popularity of green bonds has grown rapidly over the past decade, driven by the increasing demand for sustainable investments. Institutional investors, such as pension funds and insurance companies, are particularly drawn to green bonds as they align with ESG criteria. For issuers, green bonds offer an opportunity to demonstrate environmental leadership, attract a broader investor base, and often secure favorable borrowing terms.

Climate Resilience Bonds: Addressing Adaptation

Climate resilience bonds are a specialized subset of green bonds focused on projects that enhance climate adaptation and resilience. While green bonds often emphasize mitigation measures, such as reducing greenhouse gas emissions, climate resilience bonds target initiatives that help communities adapt to the impacts of climate change. These projects include flood protection, drought resilience, and coastal management systems.

For urban water management, climate resilience bonds are particularly relevant. They fund initiatives like restoring wetlands to mitigate flooding, constructing seawalls to protect against rising sea levels, and implementing stormwater systems designed to handle extreme weather events. By addressing climate adaptation, resilience bonds help cities prepare for future challenges while delivering immediate environmental and social benefits.

Key Benefits of Green and Climate Resilience Bonds

Both green bonds and climate resilience bonds offer distinct advantages for issuers, investors, and communities:

• **Diversified Capital Sources**: These bonds attract a wide range of investors, including those seeking socially responsible investment opportunities. This diversification reduces reliance on public funding and enables large-scale project implementation.

• **Transparency and Accountability**: Issuers of green and resilience bonds are required to disclose how proceeds will be used and report on the environmental impact of funded projects. This transparency builds trust among investors and stakeholders.

• **Enhanced Marketability**: Organizations issuing green or resilience bonds demonstrate a commitment to sustainability, improving their reputation and standing in the financial and public sectors.

• **Alignment with Global Goals**: These bonds directly support international frameworks such as the United Nations SDGs, particularly SDG 6 (clean water and sanitation) and SDG 13 (climate action).

Challenges and Considerations

Despite their advantages, green and climate resilience bonds face several challenges:

• **High Administrative Costs**: Issuing green bonds requires detailed impact reporting and verification, which can increase administrative burdens and costs.

• **Limited Awareness**: Some investors and potential issuers lack familiarity with these instruments, reducing their adoption in certain markets.

• **Greenwashing Risks**: Without standardized definitions and certifications, there is a risk of mislabeling bonds as "green" without ensuring they deliver genuine environmental benefits.

Efforts to address these challenges include the development of international standards, such as the Green Bond Principles (GBP), which provide guidelines for issuance and reporting. Certification frameworks, like the Climate Bonds Initiative (CBI), also play a key role in ensuring the credibility and environmental integrity of green and resilience bonds.

Realizing the Potential of Green and Climate Resilience Bonds

The future of green and climate resilience bonds lies in scaling their application to meet the growing demand for sustainable financing. Governments and financial institutions can promote these instruments by:

• **Creating Incentives**: Offering tax benefits or subsidies for green bond issuances can encourage broader participation from public and private entities.

• **Expanding Investor Education**: Raising awareness about the financial and environmental benefits of these bonds will attract more investors, particularly in emerging markets.

• **Integrating with Public Policy**: Linking green and resilience bond programs to national climate and water management policies ensures alignment with broader sustainability goals.

As global challenges like climate change and urbanization intensify, green bonds and climate resilience bonds will play an increasingly important role in financing green infrastructure. Their ability to attract diverse capital while delivering measurable environmental impact makes them indispensable tools for advancing sustainable urban water management.

Impact Investing for Green Solutions

Impact investing has emerged as a powerful financial approach that combines the pursuit of measurable environmental and social benefits with financial returns. Unlike traditional investing, which prioritizes profit maximization, impact investing seeks to generate positive outcomes for communities and ecosystems while delivering competitive returns. This dual-focus strategy has become a vital tool for financing green infrastructure projects that address urban water management challenges, climate resilience, and sustainability goals.

What Is Impact Investing?

Impact investing refers to investments made with the intention of creating positive social or environmental outcomes alongside financial returns. It is distinguished by its emphasis on measurable impact, with investors requiring evidence that their capital has achieved the desired results. In the context of green solutions, impact investing supports projects that improve water quality, reduce flood risks, and enhance urban resilience through sustainable infrastructure.

Examples of impact investing in green solutions include funding for projects like permeable pavements, urban wetlands, and rainwater harvesting systems. These investments help cities manage stormwater, adapt to climate change, and promote biodiversity while offering financial returns to investors.

The Appeal of Impact Investing

Several factors have contributed to the growing popularity of impact investing in green infrastructure:

• **Alignment with ESG Goals**: ESG criteria are increasingly important to investors, and impact investing provides a tangible way to align portfolios with these principles.

• **Market Demand for Sustainability**: Investors and consumers alike are prioritizing sustainability, creating a robust market for green infrastructure projects that deliver environmental and social benefits.

• **Long-Term Returns**: Many green infrastructure projects offer long-term financial returns through cost savings, increased property values, and ecosystem services, making them attractive to investors seeking stable and sustainable investments.

How Impact Investing Supports Green Solutions

Impact investing directly contributes to the implementation and scaling of green infrastructure projects by providing essential capital. Key mechanisms include:

• **Private Equity and Venture Capital**: Impact-focused funds invest in startups and companies developing innovative green technologies, such as smart water management systems and sustainable building materials. These investments drive technological advancements that enhance the efficiency and effectiveness of green infrastructure.

• **Debt Financing**: Investors provide loans to municipalities or private developers to fund large-scale green infrastructure projects, such as urban parks or green roofs. These loans often come with favorable terms, reflecting the social and environmental value of the projects.

• **Blended Finance**: Impact investors often collaborate with public entities or philanthropic organizations to pool resources, reducing risks and making green infrastructure projects more financially viable. For example, a municipality might provide partial funding for a rainwater harvesting initiative, while impact investors supply the remaining capital.

Measuring Impact in Green Investments

A defining feature of impact investing is the focus on measurable outcomes. Investors require clear metrics to evaluate the effectiveness of their capital in achieving environmental and social goals. For green infrastructure projects, common metrics include:

• **Stormwater Reduction**: The volume of stormwater diverted or retained through green infrastructure systems, such as bioswales or permeable pavements.

• **Water Quality Improvements**: Reductions in pollutants, such as nitrogen and phosphorus, in water bodies impacted by urban runoff.

• **Biodiversity Gains**: The creation or restoration of habitats that support wildlife in urban areas.

• **Community Benefits**: Enhanced public spaces, reduced urban heat islands, and improved air quality that benefit local residents.

Transparent reporting and third-party verification are essential for building trust with investors and ensuring accountability. Tools such as the Global Impact Investing Network's (GIIN) IRIS+ system provide standardized metrics for tracking and assessing impact across projects.

Challenges in Impact Investing

While impact investing offers significant potential, it also faces challenges:

• **Data and Reporting Gaps**: Accurately measuring and reporting impact can be complex and resource-intensive, particularly for projects with long-term or indirect benefits.

• **Perceived Risk**: Some investors view green infrastructure projects as higher-risk investments due to uncertainties in performance or regulatory changes.

• **Market Maturity**: Impact investing markets are still developing, and there may be limited opportunities or awareness in certain regions or sectors.

Addressing these challenges requires collaboration between stakeholders to improve data collection, establish clear reporting frameworks, and expand education about the financial and environmental benefits of impact investing.

The Future of Impact Investing in Green Solutions

The potential for impact investing in green infrastructure is vast, with opportunities to scale investments and drive meaningful change. As cities increasingly adopt sustainability goals, the demand for private capital to fund green solutions will continue to grow. Innovations in technology, financing models, and measurement tools will further enhance the attractiveness of impact investing, making it a cornerstone of sustainable urban development.

Governments and public entities can play a key role in fostering impact investing by creating supportive policies, offering incentives, and facilitating partnerships between investors and project developers. By aligning public and private interests, impact investing can unlock new resources to address pressing environmental challenges.

Crowdfunding and Community Financing

Crowdfunding and community financing have emerged as effective and inclusive methods for funding green infrastructure projects. These approaches leverage the collective financial contributions of individuals, businesses, and local organizations to support sustainable initiatives. By fostering grassroots engagement and empowering communities, crowdfunding and community financing enable the development of green infrastructure projects that address local water management challenges while promoting environmental awareness and social cohesion.

Crowdfunding for Green Infrastructure

Crowdfunding involves raising funds from a large number of people, typically through online platforms. This approach has gained popularity for financing smaller-scale green infrastructure projects, such as urban gardens, rainwater harvesting systems, and community green spaces.

• **How It Works**: Project organizers create campaigns on crowdfunding platforms, outlining their goals, expected benefits, and funding requirements. Individuals can contribute varying amounts, often in exchange for recognition or small rewards, such as naming rights or project updates. Platforms like Kickstarter, GoFundMe, and specialized environmental crowdfunding sites facilitate this process.

• **Advantages**: Crowdfunding democratizes project financing by allowing anyone to contribute, regardless of the amount. It also generates public interest and awareness, creating a sense of ownership and community pride in the project's success. Additionally, the visibility of crowdfunding campaigns can attract further support from local governments or private investors.

• **Challenges**: Crowdfunding relies on effective communication and marketing to reach potential donors. Projects must clearly convey their environmental and community benefits to gain traction. Additionally, the funds raised may be insufficient for larger-scale projects, limiting the scope of what can be achieved.

Community Financing Models

Community financing involves local stakeholders pooling resources to fund green infrastructure projects that directly benefit their neighborhoods. This approach emphasizes collaboration and shared responsibility, fostering stronger connections between residents, businesses, and local governments.

• **Examples of Community Financing**:

• **Cooperative Models**: Community cooperatives can finance and manage green infrastructure projects, such as urban forests or shared rainwater harvesting systems. Members contribute financially and participate in decision-making, ensuring the project aligns with local needs.

• **Local Green Funds**: Municipalities or community organizations establish dedicated funds that collect contributions from residents or businesses. These funds are used to finance small-scale projects like tree planting or bioswales, often with matching support from public agencies or grants.

• **Volunteer Contributions**: In some cases, community members contribute labor, materials, or expertise instead of money, reducing project costs and increasing community engagement.

• **Advantages**: Community financing promotes local ownership and accountability, ensuring projects are well-maintained and valued. It also fosters collaboration among diverse stakeholders, creating a sense of collective achievement.

• **Challenges**: Securing consistent participation and contributions can be difficult, especially in economically disadvantaged communities. Additionally, managing the financial and administrative aspects of community financing requires organizational capacity and coordination.

The Role of Technology in Community-Based Funding

Digital platforms and tools have played a crucial role in expanding the reach and efficiency of crowdfunding and community financing. Online platforms make it easier to launch campaigns, track contributions, and engage donors. Social media amplifies outreach efforts, allowing projects to gain visibility and attract support beyond the immediate community. Moreover, technology facilitates transparency by providing real-time updates and detailed reporting on project progress and outcomes.

Benefits of Crowdfunding and Community Financing

Both crowdfunding and community financing offer unique benefits for green infrastructure projects:

• **Increased Public Awareness**: By involving residents directly in funding and implementation, these approaches raise awareness about the importance of green infrastructure and its role in addressing urban water challenges.

• **Empowerment and Ownership**: Community involvement fosters a sense of pride and responsibility, encouraging long-term commitment to project maintenance and success.

• **Flexibility and Inclusivity**: These methods accommodate projects of varying scales and allow participation from a broad range of contributors, making green infrastructure accessible to communities with limited resources.

Overcoming Barriers to Scale

While crowdfunding and community financing are highly effective for smaller projects, scaling these models to support larger initiatives requires strategic planning and support from other stakeholders. Partnerships with local governments, private investors, or nonprofits can help bridge funding gaps and provide technical expertise. Additionally, creating regional or national frameworks for community-based financing can enhance coordination and replicate successful models in other areas.

Blending Traditional and Innovative Financing Models

The integration of traditional and innovative financing models offers a powerful approach to scaling green infrastructure projects. By combining established funding mechanisms, such as public budgets and grants, with modern tools like green bonds, impact investing, and community financing, stakeholders can maximize resources, address funding gaps, and enhance the efficiency and sustainability of project implementation.

Complementing Strengths

Traditional financing mechanisms, including government budgets, subsidies, and municipal bonds, provide a stable foundation for funding green infrastructure projects. These methods are well-suited for large-scale initiatives and ensure baseline financial support for essential urban water management systems. However, traditional models often face limitations, such as constrained public budgets and competing priorities.

Innovative financing models, such as green bonds and crowdfunding, introduce flexibility and attract diverse capital sources. They allow public agencies to leverage private investments, performance-based incentives, and community engagement, enabling the funding of cutting-edge projects that might not qualify under traditional frameworks.

By blending these approaches, stakeholders can address the weaknesses of each model. For instance, public funds can provide initial capital or guarantees for green bonds, reducing risk and attracting private investors. Similarly, community financing can be used to supplement government grants, ensuring that smaller-scale, community-driven projects receive adequate support.

Examples of Integration

PPPs: Governments collaborate with private investors and developers, combining public funds with private capital to finance and implement green infrastructure projects.

• **Hybrid Bonds**: Municipalities issue bonds that blend green and traditional elements, directing proceeds to both sustainability-focused initiatives and essential infrastructure upgrades.

• **Matching Grants**: Local governments match funds raised through community financing or crowdfunding campaigns, amplifying the impact of grassroots contributions.

3. Benefits of Blending Models

• **Diversified Funding Sources**: Combining traditional and innovative methods reduces reliance on a single funding stream, ensuring financial stability.

• **Enhanced Scalability**: Hybrid approaches make it possible to finance both large-scale and smaller, community-focused projects simultaneously.

• **Improved Stakeholder Engagement**: By involving multiple stakeholders, blended models foster collaboration and shared accountability.

Chapter 6: Performance-Based Financing for Urban Resilience

This chapter explores the role of performance-based financing in advancing green infrastructure and enhancing urban resilience. It examines the structure and benefits of key mechanisms, such as pay-for-success contracts, milestone-based payments, and outcome-based funding. Additionally, the chapter addresses the challenges associated with this approach, including data requirements, risk allocation, and stakeholder coordination.

Defining Performance-Based Financing

Performance-based financing (PBF) is an innovative approach that ties financial payments or returns to the achievement of specific, measurable outcomes. This model is designed to ensure accountability and efficiency by aligning financial incentives with project performance. In the context of urban resilience and green infrastructure, PBF focuses on delivering verifiable environmental, social, and economic benefits, such as improved stormwater management, reduced flood risks, and enhanced community well-being.

What is Performance-Based Financing?

At its core, performance-based financing shifts the traditional payment structure from input-based to results-based. Instead of funding a project solely on projected costs or completion milestones, payments are contingent upon achieving predetermined outcomes. For example, a stormwater management project might receive funding based on the volume of water retained or the reduction in pollutants entering local waterways.

This approach ensures that stakeholders are incentivized to maximize the effectiveness of their initiatives. By focusing on outcomes rather than processes, PBF encourages innovation, cost-efficiency, and the delivery of long-term benefits.

Key Features of Performance-Based Financing

Several distinguishing features define PBF:

• **Outcome-Oriented Contracts**: PBF relies on clearly defined and measurable outcomes, such as a reduction in flood events or the improvement of water quality. These metrics are established upfront and form the basis for evaluating project success.

• **Risk Sharing**: PBF involves a redistribution of risk among stakeholders. Funders or governments may share the financial risk with service providers or private investors, depending on project performance. If outcomes are not achieved, payments may be reduced or withheld entirely.

• **Independent Verification**: A critical component of PBF is the use of third-party evaluators to monitor and verify project outcomes. This independent assessment ensures transparency and builds trust among stakeholders.

• **Flexibility and Innovation**: Since payments are tied to results, service providers have greater flexibility in determining how to achieve those outcomes. This encourages creative and efficient solutions.

Types of Performance-Based Financing Models

PBF encompasses several models, each tailored to specific project needs:

• **Pay-for-Success (PFS) Contracts**: In PFS models, investors provide upfront capital for a project, and the government or other payers reimburse them based on achieved outcomes. This model is commonly used for environmental and social initiatives.

• **Milestone-Based Payments**: Payments are made at specific intervals upon achieving predetermined milestones, such as completing phases of a green infrastructure project.

• **Outcome-Based Funding**: Entire payments are contingent upon achieving final outcomes, such as reduced urban heat island effects or measurable improvements in flood resilience.

Application in Green Infrastructure and Urban Resilience

Performance-based financing is particularly well-suited to green infrastructure projects that address urban resilience challenges. For example:

• **Stormwater Management**: PBF can fund projects like rain gardens and bioswales, with payments linked to metrics such as reduced stormwater runoff volume or improved infiltration rates.

• **Flood Mitigation**: Projects aimed at reducing flood risks, such as restoring wetlands or installing permeable pavements, can be evaluated based on their effectiveness in preventing or mitigating flooding.

• **Ecosystem Services**: Urban greening projects, such as planting trees or creating green corridors, can be funded using PBF models by quantifying benefits like air quality improvement or biodiversity enhancement.

Benefits of Performance-Based Financing

PBF offers several advantages over traditional funding approaches:

• **Enhanced Accountability**: By tying payments to outcomes, PBF ensures that projects deliver tangible results and that resources are used effectively.

• **Incentivized Innovation**: Service providers and developers are motivated to adopt cost-effective and creative solutions to achieve the desired outcomes.

• **Risk Mitigation for Funders**: Governments and investors minimize financial risks by paying only for successful projects.

• **Increased Investor Confidence**: The use of independent verification builds trust, attracting socially conscious and impact-driven investors.

Challenges and Considerations

Despite its benefits, PBF presents challenges that must be addressed:

• **Defining Measurable Outcomes**: Establishing clear, quantifiable metrics can be complex, particularly for projects with long-term or indirect benefits.

• **Data and Monitoring Requirements**: PBF requires robust data collection and ongoing monitoring, which can increase administrative costs.

• **Risk Allocation**: Sharing financial risk equitably among stakeholders is essential to ensure collaboration and project success.

• **Scalability**: While effective for certain projects, scaling PBF models to larger or more complex initiatives may require additional coordination and resources.

Framework for Designing Performance-Based Contracts

Designing effective performance-based contracts (PBCs) requires a well-structured framework that aligns the interests of all stakeholders and ensures the achievement of measurable outcomes. By clearly defining roles, responsibilities, and metrics, PBCs provide a roadmap for implementing green infrastructure projects and enhancing urban resilience. The following sections outline the essential components and steps involved in creating a robust performance-based contract.

Establishing Clear Objectives

The first step in designing a performance-based contract is to define the project's goals. These objectives must be specific, measurable, achievable, relevant, and time-bound (SMART). For green infrastructure projects, objectives might include:

• Reducing stormwater runoff by a specific percentage.

• Improving water quality through pollutant filtration.

• Enhancing urban cooling by increasing green cover.

Clear objectives provide a foundation for the contract and ensure that all stakeholders have a shared understanding of the project's purpose.

Defining Performance Metrics

Performance metrics are the measurable indicators used to evaluate the project's success. These metrics must be objective, transparent, and aligned with the project's goals. Common metrics for green infrastructure projects include:

• **Stormwater Management**: Volume of stormwater retained or infiltrated.

• **Flood Risk Reduction**: Frequency or severity of flood events mitigated.

• **Water Quality Improvements**: Reductions in specific pollutants, such as nitrogen or phosphorus.

• **Ecosystem Services**: Increase in biodiversity or habitat creation.

Metrics should be selected based on their relevance to the project and the availability of data for measurement and verification.

Identifying Stakeholders and Roles

Performance-based contracts typically involve multiple stakeholders, each with distinct roles and responsibilities:

• **Issuer**: Often a government agency or municipality, the issuer sets project goals, allocates funding, and oversees implementation.

• **Service Provider**: The entity responsible for designing, implementing, and managing the project. This could include contractors, engineers, or environmental organizations.

• **Investors**: In some cases, private investors provide upfront capital, with repayments tied to the achievement of outcomes.

• **Independent Evaluator**: A third-party organization tasked with monitoring and verifying project performance against the established metrics.

Clearly defining stakeholder roles helps ensure accountability and smooth project execution.

Structuring Financial Incentives

A critical feature of performance-based contracts is the alignment of financial incentives with project outcomes. Payments are tied to the achievement of predefined performance metrics, creating a results-driven model. Typical structures include:

• **Pay-for-Success**: Payments are made only when the project meets or exceeds specific targets, such as reducing stormwater runoff by a set amount.

• **Milestone-Based Payments**: Partial payments are made at various stages of the project, contingent on achieving interim goals.

• **Tiered Incentives**: Bonus payments or penalties based on performance levels, incentivizing service providers to exceed baseline expectations.

These financial structures motivate stakeholders to prioritize efficiency and innovation while managing risks.

Establishing Monitoring and Verification Processes

Effective monitoring and verification are essential for evaluating performance and ensuring transparency. A robust process includes:

• **Baseline Data Collection**: Establishing initial conditions to provide a reference point for measuring progress.

• **Regular Monitoring**: Ongoing collection of data to track project performance over time.

• **Third-Party Evaluation**: Independent verification of outcomes to ensure accuracy and build trust among stakeholders.

Technological tools, such as sensors, remote monitoring systems, and Geographic Information Systems (GIS), can enhance data collection and reporting.

Allocating Risks and Responsibilities

Risk allocation is a critical aspect of performance-based contracts. Financial, operational, and performance risks must be distributed equitably among stakeholders. For example:

• The service provider may bear the risk of underperformance if payments are contingent on achieving outcomes.

• The issuer may assume responsibility for external risks, such as regulatory changes or extreme weather events.

Clearly defining risk-sharing arrangements in the contract ensures that stakeholders remain committed to the project's success.

Including Dispute Resolution Mechanisms

Given the complexity of performance-based contracts, disputes may arise during project implementation. Including clear dispute resolution mechanisms, such as mediation or arbitration clauses, helps prevent delays and ensures that conflicts are resolved efficiently.

Building Flexibility into the Contract

Green infrastructure projects often involve dynamic environmental conditions and long timelines. Performance-based contracts should include provisions for flexibility, allowing stakeholders to adjust metrics or timelines in response to unforeseen challenges or new information.

Outcome-Based Metrics for Urban Resilience

Outcome-based metrics are essential for evaluating the success of urban resilience initiatives, particularly those involving green infrastructure. These metrics focus on measurable results that demonstrate the effectiveness of projects in addressing urban water management challenges, enhancing climate adaptation, and improving environmental quality. By providing a clear and transparent framework for monitoring and evaluation, outcome-based metrics ensure accountability and drive performance-based financing models.

Importance of Outcome-Based Metrics

Outcome-based metrics are designed to measure the tangible results of green infrastructure projects, distinguishing them from input- or output-based approaches. Instead of focusing on activities (e.g., the number of trees planted), outcome-based metrics assess the direct benefits achieved (e.g., reduction in stormwater runoff or urban heat). These metrics are critical for several reasons:

• **Accountability**: They tie financial incentives and stakeholder responsibilities to verifiable results, ensuring that resources are used efficiently.

• **Decision-Making**: Metrics provide data-driven insights for planning, funding, and scaling green infrastructure projects.

• **Transparency**: Clear outcomes build trust among stakeholders, including investors, governments, and communities.

• **Adaptability**: Metrics help identify areas for improvement, enabling project adjustments based on real-time data.

Key Outcome-Based Metrics for Urban Resilience

Metrics for urban resilience projects are tailored to the goals of green infrastructure and the specific challenges being addressed. Common categories include:

• **Stormwater Management**:

• **Runoff Reduction**: Volume of stormwater captured or infiltrated through systems like bioswales, rain gardens, or permeable pavements.

• **Peak Flow Mitigation**: Reduction in peak flow rates during heavy rainfall, preventing overloading of drainage systems.

• **Pollutant Removal**: Amount of contaminants, such as nitrogen, phosphorus, and suspended solids, removed from stormwater before it enters waterways.

• **Flood Risk Reduction**:

• **Flood Event Frequency**: Decrease in the occurrence of flood events in urban areas.

• **Flood Damage Costs**: Reduction in financial losses attributed to flooding, including property damage and infrastructure repair costs.

• **Floodplain Restoration**: Area of restored floodplains contributing to natural water retention and ecosystem services.

• **Climate Adaptation**:

• **Urban Cooling**: Reduction in local temperatures due to increased tree canopy cover or green roofs, mitigating urban heat island effects.

• **Drought Resilience**: Increased water retention and storage capacity, reducing reliance on external water sources during dry periods.

• **Ecosystem Services**: Improvements in biodiversity, air quality, and carbon sequestration provided by green infrastructure.

• **Community and Social Benefits**:

• **Public Space Utilization**: Increase in the number of people accessing and using green spaces created through urban resilience projects.

• **Health and Well-Being**: Reduction in heat-related illnesses, respiratory conditions, or stress due to improved environmental conditions.

• **Equity**: Distribution of project benefits across socioeconomic and geographic boundaries, ensuring support for underserved communities.

Methods for Measuring Metrics

Accurate measurement of outcome-based metrics requires robust data collection and evaluation methodologies. Key methods include:

• **Baseline Assessments**: Establishing initial conditions to provide a reference point for evaluating project impact. For example, baseline measurements of runoff volumes or water quality can be compared to post-implementation results.

• **Monitoring Systems**: Utilizing sensors, remote sensing technology, and GIS to collect real-time data on stormwater, temperature, or other environmental indicators.

• **Community Surveys**: Engaging local residents to gather qualitative data on social benefits, such as improved access to green spaces or perceived environmental improvements.

• **Third-Party Verification**: Independent evaluators ensure the accuracy and credibility of reported outcomes, particularly for performance-based financing models.

Challenges in Defining and Measuring Metrics

Despite their importance, outcome-based metrics present several challenges:

• **Data Availability**: Reliable and consistent data can be difficult to obtain, particularly in resource-limited settings or for long-term impacts.

• **Attribution**: Isolating the effects of a single project from broader environmental or social changes can be complex.

• **Metric Selection**: Identifying the most relevant and impactful metrics requires careful planning and alignment with project goals.

• **Cost of Monitoring**: Implementing comprehensive data collection and evaluation systems can increase project costs.

Best Practices for Implementing Outcome-Based Metrics

To overcome challenges and maximize the effectiveness of metrics, stakeholders should consider the following best practices:

• **Stakeholder Collaboration**: Involve all relevant parties, including funders, service providers, and communities, in defining metrics and establishing monitoring protocols.

• **Standardization**: Use established frameworks, such as the GIIN IRIS+ system, to standardize metrics and reporting.

• **Adaptive Management**: Regularly review metrics and adjust project strategies based on findings to ensure continuous improvement.

Scaling Performance-Based Financing Globally

Performance-based financing (PBF) has proven to be a transformative approach for funding green infrastructure and enhancing urban resilience. By linking financial returns to measurable outcomes, this model incentivizes efficiency, innovation, and accountability. As cities worldwide face increasing challenges from climate change, urbanization, and water management issues, scaling PBF globally offers a pathway to achieve sustainable solutions on a broader scale.

Global Demand for Scalable Solutions

The rising impacts of climate change, such as intensified flooding, droughts, and heatwaves, are driving demand for resilient infrastructure in cities across the globe. Traditional financing mechanisms alone are insufficient to meet this growing need. PBF offers a scalable solution that attracts diverse funding sources, including private investors, international organizations, and philanthropic institutions. Its ability to align financial incentives with environmental outcomes makes it particularly suited for addressing global challenges in urban water management.

Leveraging International Collaboration

Scaling PBF globally requires fostering collaboration between governments, financial institutions, and international organizations. Entities such as the United Nations, the World Bank, and regional development banks play a critical role in supporting PBF initiatives through funding, technical assistance, and knowledge-sharing platforms. By facilitating partnerships and providing capacity-building resources, these organizations can help lower barriers to adoption in developing countries and emerging markets.

Standardizing Metrics and Frameworks

A major challenge in scaling PBF globally is the lack of consistent metrics and frameworks for evaluating project outcomes. Establishing international standards for outcome-based metrics, monitoring protocols, and reporting processes is essential to build trust among stakeholders and attract investment. Initiatives such as the Global Impact Investing Network (GIIN) and the Climate Bonds Initiative (CBI) are already contributing to the development of standardized tools that enhance transparency and accountability.

Expanding Technological Integration

Advancements in technology can significantly enhance the scalability of PBF. Tools such as remote sensing, GIS, and IoT devices enable efficient data collection, real-time monitoring, and outcome verification. By integrating these technologies into PBF models, cities can reduce administrative costs, improve performance tracking, and ensure project success on a global scale.

Promoting Policy and Regulatory Support

Governments have a vital role in creating an enabling environment for scaling PBF. Policies that incentivize private investment, streamline permitting processes, and integrate PBF into national climate and water management strategies can encourage adoption. Regulatory frameworks that support outcome-based financing ensure that projects align with long-term sustainability goals.

Chapter 7: Integrating Green Infrastructure into Urban Planning and Policy

This chapter explores how cities can incorporate green infrastructure into their planning and policy frameworks. It examines key strategies, such as updating zoning laws to encourage permeable surfaces, requiring green roofs in new developments, and aligning green infrastructure initiatives with climate adaptation plans. The chapter also addresses the challenges of integration, including funding constraints, stakeholder coordination, and balancing competing land-use priorities.

Mainstreaming Green Infrastructure in Urban Planning

Green infrastructure is no longer a niche concept but a critical component of sustainable urban development. By managing stormwater, mitigating urban heat islands, and enhancing biodiversity, green infrastructure provides multifunctional solutions to some of the most pressing challenges faced by cities. Mainstreaming green infrastructure in urban planning requires its integration into policies, regulations, and practices that guide city development, ensuring that it becomes a core element of urban resilience and sustainability.

Importance of Mainstreaming Green Infrastructure

Mainstreaming green infrastructure means embedding it into the broader urban planning framework, making it a standard practice rather than an exception. This approach recognizes that green infrastructure is essential for achieving long-term environmental, social, and economic benefits. Key reasons for mainstreaming include:

• **Climate Resilience**: Green infrastructure reduces vulnerabilities to climate-related risks, such as flooding, drought, and extreme heat, by providing natural solutions that complement traditional infrastructure.

• **Cost-Effectiveness**: By addressing multiple urban challenges simultaneously, green infrastructure often proves more cost-effective than conventional gray infrastructure over the long term.

• **Livability and Equity**: Integrating green infrastructure improves urban livability by creating green spaces, enhancing air quality, and providing recreational opportunities, particularly in underserved communities.

Strategies for Mainstreaming Green Infrastructure

To ensure that green infrastructure becomes a fundamental part of urban planning, cities can adopt the following strategies:

• **Incorporating Green Infrastructure in Master Plans**: Urban master plans should explicitly include green infrastructure as a priority, outlining specific goals and targets for its implementation. This integration ensures that green infrastructure is considered in all stages of urban development, from design to execution.

• **Updating Zoning and Building Codes**: Zoning regulations and building codes can be revised to require or incentivize green infrastructure. For example, zoning ordinances might mandate permeable pavements in parking lots or green roofs on commercial buildings. Similarly, density bonuses can encourage developers to include features like bioswales or rain gardens in their projects.

• **Establishing Dedicated Policies and Programs**: Cities can develop policies that focus exclusively on green infrastructure, such as stormwater management plans or urban greening initiatives.

Dedicated programs provide clear guidance for stakeholders and ensure that resources are allocated to meet green infrastructure goals.

• **Integrating Green Infrastructure with Climate Plans**: Linking green infrastructure to climate adaptation and mitigation plans strengthens its role in addressing urban climate challenges. For instance, urban heat island reduction strategies can prioritize tree planting and vegetative cover as critical components.

Tools for Implementation

To successfully mainstream green infrastructure, urban planners and policymakers need access to effective tools and resources, including:

• **Geospatial Analysis and Mapping**: Tools like GIS can identify areas most in need of green infrastructure interventions, such as flood-prone zones or heat-vulnerable neighborhoods.

• **Community Engagement Platforms**: Engaging communities in the planning and implementation of green infrastructure fosters local support and ensures that projects address specific community needs.

• **Performance Metrics and Monitoring**: Establishing measurable indicators, such as stormwater retention or urban cooling, helps evaluate the effectiveness of green infrastructure initiatives and informs future planning.

Challenges in Mainstreaming Green Infrastructure

Despite its benefits, integrating green infrastructure into urban planning faces several challenges:

• **Funding Constraints**: Limited budgets often prioritize traditional infrastructure over green solutions. Identifying innovative financing models, such as green bonds or public-private partnerships, can address this barrier.

• **Stakeholder Coordination**: Collaboration among multiple stakeholders, including governments, developers, and community groups, is essential but can be complex.

• **Competing Land-Use Priorities**: Urban areas often face pressure to maximize land use for housing or commercial development, leaving limited space for green infrastructure.

• **Knowledge Gaps**: Planners and policymakers may lack technical knowledge about green infrastructure, underscoring the need for training and capacity-building programs.

Case for Mainstreaming Green Infrastructure

By mainstreaming green infrastructure, cities can create urban environments that are more resilient, equitable, and sustainable. For example, integrating green infrastructure into redevelopment projects can transform underutilized spaces into vibrant community assets. Similarly, prioritizing green infrastructure in flood-prone areas can protect vulnerable populations while enhancing biodiversity and recreational opportunities.

Regulatory Frameworks and Policy Alignment

Integrating green infrastructure into urban development requires the support of robust regulatory frameworks and aligned policies. These tools establish the legal, institutional, and operational foundations necessary to prioritize, implement, and maintain green infrastructure solutions. By aligning green infrastructure initiatives with broader urban policies, governments can drive widespread adoption, foster collaboration among stakeholders, and ensure long-term sustainability.

Importance of Regulatory Frameworks

Regulatory frameworks provide the rules and standards that guide the planning, design, and implementation of green infrastructure. They ensure that projects meet environmental, social, and economic objectives while addressing urban challenges such as stormwater management, climate adaptation, and biodiversity loss. Effective regulatory frameworks:

• **Promote Consistency**: Clear regulations ensure that green infrastructure standards are consistently applied across projects and regions.

• **Encourage Compliance**: Legal mandates and incentives motivate developers and property owners to adopt sustainable practices.

• **Support Innovation**: Regulations that accommodate new technologies and methods enable creative solutions to urban water management challenges.

Key Elements of Effective Regulatory Frameworks

Effective regulatory frameworks for green infrastructure typically include the following elements:

• **Clear Standards and Guidelines**: Detailed technical standards define the design, construction, and maintenance of green infrastructure features. For example, stormwater management regulations may specify runoff reduction targets or design criteria for bioswales and permeable pavements.

• **Mandates for Implementation**: Legal requirements, such as zoning ordinances or building codes, mandate the inclusion of green infrastructure in specific developments or areas. For instance, cities may require green roofs on new commercial buildings or the use of permeable materials in parking lots.

• **Monitoring and Enforcement Mechanisms**: Regulations must include processes for monitoring compliance and enforcing standards. Regular inspections and penalties for non-compliance ensure accountability and maintain project integrity.

Policy Alignment for Green Infrastructure

Aligning green infrastructure with broader urban policies strengthens its integration into city planning and development. This alignment involves embedding green solutions into policies related to land use, climate adaptation, public health, and economic development.

• **Land-Use Policies**: Integrating green infrastructure into zoning laws and land-use plans ensures that sustainable practices are prioritized in urban development. For example, zoning ordinances can designate areas for green corridors, urban forests, or floodplain restoration.

• **Climate Adaptation Strategies**: Green infrastructure aligns naturally with climate adaptation policies, such as those addressing flood mitigation, urban cooling, and water resource management. Policies that link green infrastructure to national or local climate goals amplify its role in building resilience.

• **Public Health Initiatives**: Green infrastructure contributes to public health by improving air quality, reducing urban heat, and creating recreational spaces. Policies that promote these benefits strengthen the case for green infrastructure investment.

• **Economic Development Plans**: Aligning green infrastructure with economic policies can attract investments and create green jobs. For example, policies that incentivize green construction methods support both environmental and economic goals.

Incentives for Compliance

Regulatory frameworks often incorporate incentives to encourage voluntary compliance and accelerate adoption. Common incentives include:

• **Tax Benefits**: Property tax reductions or credits for developers who incorporate green infrastructure into their projects.

• **Fee Reductions**: Stormwater fee discounts for property owners who implement runoff-reducing measures, such as rain gardens or cisterns.

• **Grants and Subsidies**: Financial support for installing and maintaining green infrastructure features, particularly for small businesses or low-income communities.

• **Expedited Permitting**: Fast-tracked approval processes for developments that meet green infrastructure standards.

These incentives reduce the financial and administrative barriers associated with green infrastructure, making it a more viable option for developers and property owners.

Addressing Challenges in Policy and Regulation

Despite their importance, regulatory frameworks and policy alignment face several challenges:

• **Complexity and Overlap**: Conflicting or overlapping regulations can create confusion and inefficiencies. Streamlining policies across government agencies is essential to avoid redundancies and inconsistencies.

• **Enforcement Gaps**: Limited resources for monitoring and enforcement can undermine compliance, particularly in cities with high development pressures.

• **Resistance to Change**: Developers and property owners may resist new regulations due to perceived costs or administrative burdens. Outreach and education can help address these concerns and highlight the benefits of green infrastructure.

• **Equity Considerations**: Policies must ensure that the benefits of green infrastructure are equitably distributed across all communities, particularly those historically underserved.

International Examples of Regulatory Success

Many cities and countries have implemented successful regulatory frameworks for green infrastructure:

• **Singapore**: The "ABC Waters Programme" integrates green infrastructure into urban planning, with regulatory requirements for stormwater management and water-sensitive urban design.

• **Germany**: Cities like Berlin mandate green roofs on new developments, supported by clear technical guidelines and financial incentives.

• **United States**: The Clean Water Act provides a regulatory foundation for green infrastructure, with cities like Philadelphia implementing stormwater fees and credits to incentivize sustainable practices.

Global Sustainability Goals and Local Strategies

The integration of green infrastructure into urban planning is crucial for achieving global sustainability goals. Initiatives such as the United Nations SDGs and the Paris Agreement provide a global framework for addressing climate change, water management, and urban resilience. However, the successful implementation of these goals depends on translating them into actionable local strategies that reflect the unique challenges and opportunities of individual communities.

Aligning Green Infrastructure with Global Goals

Green infrastructure directly supports several global sustainability goals by addressing key environmental, social, and economic priorities:

• **SDG 6 (Clean Water and Sanitation)**: Green infrastructure solutions, such as rain gardens and bioswales, improve water quality by reducing pollutants in stormwater runoff and enhancing natural filtration processes.

• **SDG 11 (Sustainable Cities and Communities)**: By integrating green spaces and sustainable water management systems, green infrastructure contributes to more inclusive, safe, and resilient urban environments.

• **SDG 13 (Climate Action)**: Green infrastructure helps mitigate the effects of climate change by managing flood risks, reducing urban heat islands, and supporting carbon sequestration.

In addition to the SDGs, the Paris Agreement emphasizes the importance of adaptation measures, many of which can be addressed through green infrastructure. For example, restoring wetlands and implementing permeable pavements reduce vulnerabilities to extreme weather events while contributing to long-term urban resilience.

Translating Global Goals into Local Strategies

While global frameworks set the direction, local strategies are essential for implementing green infrastructure effectively. These strategies must consider the specific environmental conditions, economic resources, and social dynamics of each community.

• **Localized Climate Adaptation Plans**: Cities can integrate green infrastructure into their climate adaptation strategies by identifying high-risk areas for flooding, drought, or heat stress. For example, urban neighborhoods with limited green spaces may prioritize tree planting and vegetative barriers to reduce temperatures and improve air quality.

• **Community-Driven Planning**: Engaging local communities in the design and implementation of green infrastructure projects ensures that strategies address their unique needs and priorities. For instance, community consultations can guide the placement of rain gardens or urban wetlands to maximize accessibility and impact.

• **Partnership Development**: Collaborations between local governments, private developers, and nonprofit organizations enable the pooling of resources and expertise for green infrastructure projects. These partnerships can also align local efforts with broader regional or national sustainability goals.

Challenges in Bridging Global and Local Efforts

Implementing global sustainability goals at the local level presents several challenges:

• **Resource Limitations**: Many cities, particularly in developing countries, face financial and technical constraints that hinder the adoption of green infrastructure.

• **Policy Misalignment**: Inconsistent or conflicting policies at different levels of government can create barriers to effective implementation. For example, national priorities may not always align with local needs.

• **Capacity Gaps**: Local governments and communities may lack the expertise or training needed to design, implement, and maintain green infrastructure projects.

Addressing these challenges requires targeted investments in capacity-building, policy coordination, and innovative financing mechanisms, such as public-private partnerships or green bonds.

Examples of Localized Approaches to Global Goals

Several cities have successfully translated global sustainability goals into actionable local strategies:

• **Copenhagen, Denmark**: The city's Cloudburst Management Plan aligns with global climate goals by implementing green infrastructure to manage extreme rainfall events, including permeable pavements and urban retention ponds.

• **Cape Town, South Africa**: Local water conservation initiatives, such as the use of green roofs and rainwater harvesting systems, support SDG 6 by addressing water scarcity.

• **New York City, USA**: The Green Infrastructure Program reduces stormwater runoff while contributing to SDG 11 by enhancing public green spaces and improving urban resilience.

Role of Stakeholder Collaboration

Stakeholder collaboration is a cornerstone of successfully implementing green infrastructure projects and advancing sustainable urban development. By fostering partnerships among diverse groups, including government agencies, private sector actors, nonprofit organizations, and local communities, cities can overcome challenges, pool resources, and maximize the benefits of green infrastructure. Collaborative approaches ensure that projects are well-informed, widely supported, and effectively executed.

Importance of Stakeholder Collaboration

Green infrastructure projects are inherently interdisciplinary, requiring input and coordination from various stakeholders. Collaboration is vital for several reasons:

• **Shared Expertise**: Different stakeholders bring unique knowledge and skills. For example, engineers provide technical insights, urban planners contribute design expertise, and local communities offer practical knowledge of the area's needs and challenges.

• **Resource Mobilization**: Collaboration enables stakeholders to pool financial, technical, and human resources, reducing the burden on any single entity and enhancing the feasibility of projects.

• **Broad Support**: Engaging diverse stakeholders fosters buy-in and minimizes opposition. Inclusive planning processes ensure that projects address the needs of all affected groups, increasing their chances of success.

Key Stakeholders in Green Infrastructure

Collaboration in green infrastructure typically involves the following stakeholders:

• **Government Agencies**: Municipal, regional, and national governments play a leading role in setting policies, providing funding, and overseeing project implementation.

• **Private Sector**: Developers, investors, and businesses contribute financial capital, innovation, and operational expertise. PPPs are particularly effective in scaling green infrastructure initiatives.

• **Nonprofit Organizations and Advocacy Groups**: These organizations often act as intermediaries, facilitating communication between governments and communities and advocating for equitable and sustainable practices.

• **Local Communities**: Community involvement ensures that projects address specific local needs and gain public support. Residents may also contribute labor, materials, or maintenance efforts.

• **Academia and Research Institutions**: These groups provide data, tools, and frameworks for designing and evaluating green infrastructure projects, contributing to evidence-based decision-making.

Benefits of Collaboration

Collaboration among stakeholders offers several advantages:

• **Innovation and Creativity**: Bringing together diverse perspectives fosters innovative solutions that address complex urban challenges.

• **Efficiency and Cost-Effectiveness**: Shared resources and responsibilities reduce duplication of efforts and optimize project outcomes.

• **Long-Term Sustainability**: Stakeholder engagement in planning and implementation promotes accountability and ensures that green infrastructure is maintained over time.

Challenges and Strategies for Effective Collaboration

Despite its benefits, stakeholder collaboration can face obstacles such as conflicting priorities, communication gaps, and resource limitations. Strategies to enhance collaboration include:

• **Clear Roles and Responsibilities**: Clearly defining stakeholder roles reduces confusion and ensures accountability.

• **Open Communication**: Regular meetings, updates, and feedback loops build trust and facilitate problem-solving.

• **Equity and Inclusion**: Ensuring that all voices, particularly those of underserved communities, are heard and valued fosters more inclusive and equitable outcomes.

Chapter 8: Overcoming Financial Barriers to Green Infrastructure

Green infrastructure offers transformative solutions for urban water management, climate resilience, and environmental sustainability. However, one of the most significant challenges to its widespread implementation is overcoming financial barriers. Despite its long-term cost-effectiveness and multiple co-benefits, the upfront costs of planning, designing, and constructing green infrastructure can deter public agencies, private developers, and communities from pursuing these projects. Furthermore, maintenance and monitoring expenses can place additional financial burdens on already constrained budgets.

Traditional funding mechanisms often prioritize gray infrastructure due to its perceived reliability and familiarity. At the same time, green infrastructure projects may struggle to attract investment because of uncertainties around long-term performance, data availability, and the complexity of quantifying their benefits. These challenges highlight the need for innovative financial solutions and collaborative approaches to unlock the full potential of green infrastructure.

This chapter examines the financial barriers hindering the adoption of green infrastructure and explores strategies to overcome them. It discusses innovative funding mechanisms, such as green bonds and public-private partnerships, as well as the importance of aligning financial planning with sustainability goals. The chapter also highlights the role of government incentives, private investment, and community-based financing in bridging funding gaps.

By addressing these barriers, cities and stakeholders can create a supportive financial environment for green infrastructure, ensuring its integration into urban planning and fostering long-term sustainability. This chapter provides practical insights for overcoming the financial challenges that often limit the implementation of these essential projects.

Identifying Financial Challenges

Implementing green infrastructure on a large scale presents significant financial challenges. While green solutions offer long-term environmental, social, and economic benefits, the upfront costs, ongoing maintenance requirements, and funding uncertainties can create barriers for public agencies, private developers, and communities. Understanding these financial challenges is critical to developing effective strategies for overcoming them and ensuring the successful adoption of green infrastructure.

High Initial Costs

One of the most significant financial challenges is the high upfront cost of planning, designing, and constructing green infrastructure projects. Features like green roofs, permeable pavements, and rain gardens often require specialized materials, expertise, and site preparation, which can lead to higher initial expenditures compared to traditional gray infrastructure. For example:

• **Design and Engineering**: Creating tailored green infrastructure solutions requires specialized knowledge and technical expertise, increasing upfront design costs.

• **Construction Materials**: Sustainable materials, such as permeable concrete or native plants, may be more expensive or less readily available than conventional alternatives.

• **Land Acquisition**: In densely populated urban areas, securing land for green infrastructure can be costly, particularly in high-value zones.

These initial costs can discourage investment, particularly in cities with constrained budgets or competing infrastructure priorities.

Ongoing Maintenance Expenses

Green infrastructure requires regular maintenance to ensure its effectiveness and longevity. Unlike traditional infrastructure, which may have fewer upkeep requirements, green solutions rely on continuous care, such as:

• **Landscaping**: Regular pruning, weeding, and replanting of vegetation.

• **Stormwater System Cleaning**: Clearing sediment and debris from bioswales, permeable pavements, and other systems to maintain performance.

• **Monitoring and Repairs**: Using sensors and inspections to track performance and address wear and tear over time.

For municipalities and property owners, these ongoing expenses can strain budgets, especially if maintenance costs were not adequately planned during the project's initial phase.

Limited Access to Funding

Access to reliable funding is a common challenge for green infrastructure projects, as traditional financing mechanisms often prioritize gray infrastructure. Green infrastructure may struggle to compete for public funds due to:

• **Budget Constraints**: Municipal budgets are frequently limited, and resources are often directed toward critical needs like education, healthcare, and traditional infrastructure.

• **Short-Term Focus**: Decision-makers may prioritize projects with immediate, tangible results, overlooking the long-term benefits of green infrastructure.

• **Competing Priorities**: Funding may be diverted to other urban development initiatives, leaving green infrastructure underfunded.

Private sector financing can also be difficult to secure, particularly when investors perceive green infrastructure as a high-risk or low-return investment.

Complexity of Quantifying Benefits

Green infrastructure delivers a range of co-benefits, such as improved water quality, enhanced biodiversity, and reduced urban heat islands. However, these benefits can be difficult to quantify, making it challenging to demonstrate the value of green infrastructure to funders and investors. Common issues include:

• **Lack of Data**: Limited availability of baseline data or performance metrics for green infrastructure projects.

• **Intangible Benefits**: Difficulty in assigning monetary value to ecosystem services, public health improvements, or community well-being.

• **Long-Term Outcomes**: Benefits such as flood risk reduction or climate adaptation may take years to materialize, complicating cost-benefit analyses.

Without clear and quantifiable evidence of green infrastructure's value, projects may struggle to attract funding from both public and private sources.

Inadequate Policy and Regulatory Support

Financial barriers are often compounded by policy and regulatory challenges that hinder green infrastructure adoption. These challenges include:

• **Lack of Incentives**: Insufficient financial incentives, such as tax credits or grants, to encourage investment in green solutions.

• **Outdated Regulations**: Building codes and zoning laws that prioritize gray infrastructure over sustainable alternatives.

• **Fragmented Governance**: Limited coordination between agencies or levels of government, resulting in inconsistent funding priorities.

These gaps in policy and regulation can discourage stakeholders from pursuing green infrastructure projects, exacerbating financial challenges.

Strategies for Cost Reduction and Resource Scaling

Implementing green infrastructure at scale requires strategic approaches to reduce costs and optimize resources. While green infrastructure offers long-term environmental and social benefits, upfront costs, ongoing maintenance, and funding gaps often act as barriers. By leveraging innovative strategies and collaborative approaches, stakeholders can overcome these challenges and create cost-effective, scalable solutions that integrate seamlessly into urban environments.

Efficient Design and Planning

Cost reduction begins with careful design and planning that optimize the use of resources. By tailoring solutions to local conditions and needs, stakeholders can minimize unnecessary expenses while maximizing project impact.

• **Site-Specific Solutions**: Designing green infrastructure to address specific local challenges, such as flood-prone areas or heat islands, ensures resources are allocated effectively. For example, choosing native plants reduces maintenance costs and improves ecosystem integration.

• **Multi-Functional Design**: Incorporating multiple functions into a single project, such as stormwater management, urban cooling, and recreational space, enhances the cost-effectiveness of green infrastructure investments.

• **Phased Implementation**: Rolling out projects in phases allows stakeholders to manage costs incrementally and adapt designs based on lessons learned from earlier phases.

Leveraging Economies of Scale

Scaling green infrastructure across larger areas or multiple projects can reduce costs through economies of scale. Bulk purchasing, standardized designs, and coordinated implementation help lower overall expenses.

• **Bulk Procurement**: Purchasing materials, such as permeable pavements or plants, in large quantities reduces unit costs, making projects more affordable.

• **Standardized Components**: Using standardized designs for commonly implemented green infrastructure elements, such as rain gardens or green roofs, streamlines construction and reduces engineering costs.

• **Regional Collaboration**: Partnering with neighboring municipalities to implement regional green infrastructure projects fosters resource sharing and cost savings.

Innovative Financing Mechanisms

Accessing diverse and innovative financing mechanisms enables stakeholders to reduce reliance on traditional funding sources and scale green infrastructure more effectively.

• **PPPs**: Collaborations between governments and private sector entities allow for cost-sharing, with private partners often contributing technical expertise and upfront capital.

• **Green Bonds**: Issuing bonds specifically tied to sustainable projects provides cities with access to low-cost capital while appealing to environmentally conscious investors.

• **Performance-Based Financing**: Tying payments to measurable outcomes, such as reduced stormwater runoff or improved water quality, ensures that resources are used efficiently and incentivizes high performance.

Community Engagement and Resource Mobilization

Engaging local communities in green infrastructure projects not only fosters public support but also reduces costs by mobilizing additional resources.

• **Volunteer Participation**: Encouraging community members to contribute labor or materials for projects, such as tree planting or rain garden installation, reduces upfront costs.

• **Crowdfunding**: Platforms like GoFundMe or Kickstarter enable communities to raise small contributions from a large number of donors, filling funding gaps for smaller-scale projects.

• **In-Kind Contributions**: Partnering with local businesses or organizations to provide materials, expertise, or services can significantly lower costs.

Integrating Green Infrastructure with Existing Projects

Aligning green infrastructure initiatives with ongoing or planned urban development projects reduces costs by leveraging existing budgets and resources.

• **Retrofitting Existing Infrastructure**: Converting underutilized spaces, such as vacant lots or rooftops, into green infrastructure features minimizes land acquisition costs and repurposes existing assets.

• **Co-Locating Infrastructure**: Combining green infrastructure with gray infrastructure projects, such as adding bioswales alongside roadways, maximizes efficiency and reduces duplication of efforts.

• **Incorporating Green Solutions in New Developments**: Mandating green infrastructure in new developments through updated zoning regulations or building codes ensures that costs are incorporated into project budgets from the outset.

Policy Support and Incentives

Government policies and incentives play a crucial role in reducing costs and scaling green infrastructure projects.

• **Tax Incentives and Subsidies**: Offering property tax reductions, grants, or rebates for implementing green infrastructure encourages investment while offsetting upfront costs for developers and property owners.

• **Fee Reductions**: Stormwater fee discounts for properties that install features like rain gardens or permeable pavements incentivize adoption and reduce long-term costs for municipalities.

• **Streamlined Permitting**: Simplifying permitting processes for green infrastructure projects reduces administrative burdens and accelerates implementation timelines.

Advancing Technological Innovation

Technological advancements can significantly reduce costs and improve resource efficiency in green infrastructure implementation.

• **Smart Monitoring Systems**: Using IoT devices or sensors for real-time monitoring of stormwater systems minimizes labor-intensive inspections and maintenance.

• **Modular Components**: Prefabricated modular systems, such as green roof panels or pre-designed bioswale units, reduce installation time and labor costs.

• **GIS and Modeling Tools**: GIS and simulation tools aid in planning and designing efficient green infrastructure layouts, optimizing resource allocation.

Long-Term Cost Management

To ensure cost-effectiveness over time, stakeholders must prioritize long-term maintenance and sustainability.

• **Public-Private Maintenance Agreements**: Establishing contracts that outline shared responsibilities for ongoing maintenance reduces financial burdens on any single entity.

• **Maintenance Training Programs**: Equipping local workers or community members with skills to maintain green infrastructure reduces dependency on external contractors.

• **Adaptive Management**: Regular evaluations and adjustments to maintenance plans based on performance data ensure that resources are allocated efficiently.

Enabling Innovation in Low-Resource Settings

Implementing green infrastructure in low-resource settings requires innovative approaches that overcome financial, technical, and institutional barriers. By leveraging locally available materials, fostering community engagement, and applying cost-effective solutions, cities and communities with limited resources can adopt sustainable practices that address urban challenges like stormwater management and climate resilience. Enabling innovation in these contexts ensures that green infrastructure is not only a solution for well-funded urban centers but also a viable option for underserved and developing areas.

Leveraging Locally Available Resources

A key strategy for implementing green infrastructure in low-resource settings is utilizing materials and knowledge that are readily available within the community. This approach reduces costs while fostering local ownership and adaptability.

• **Local Materials**: Incorporating native plants, locally sourced soil, and recycled materials minimizes the need for costly imports. For example, recycled bricks can be repurposed for permeable pavements, and local vegetation can be used in rain gardens.

• **Traditional Knowledge**: Drawing on indigenous or traditional practices for water management and land use provides cost-effective and culturally relevant solutions. Communities with a history of managing water resources can contribute valuable insights into designing and maintaining green infrastructure.

• **Low-Cost Designs**: Simplifying designs to focus on essential functions reduces complexity and associated expenses. For example, basic bioswales or retention ponds can be constructed without requiring advanced technology.

Community Engagement and Participation

Community involvement is critical for the success of green infrastructure projects in low-resource settings. Engaging local residents ensures that projects address specific needs and encourages active participation in their implementation and maintenance.

• **Volunteer Contributions**: Communities can provide labor for tasks such as planting vegetation, constructing simple structures, or maintaining green spaces. This reduces costs while fostering a sense of collective ownership.

• **Education and Training**: Providing training programs equips residents with the skills to design, implement, and maintain green infrastructure projects. These programs build local capacity and create job opportunities.

• **Collaborative Planning**: Involving community members in the planning process ensures that projects align with local priorities and constraints. For example, residents can identify areas prone to flooding or locations suitable for green space development.

Innovative Financing Mechanisms

Securing funding is a major challenge in low-resource settings, but innovative financing models can help address this barrier.

• **Microfinance and Crowdfunding**: Small-scale loans or community fundraising campaigns provide accessible financing options for localized green infrastructure projects. For example, crowdfunding platforms can support initiatives like installing rainwater harvesting systems.

• **PPPs**: Collaborations between local governments, NGOs, and private businesses enable resource pooling and shared responsibilities, reducing the financial burden on individual entities.

• **In-Kind Contributions**: Partnerships with local businesses can provide materials, tools, or technical expertise at reduced or no cost, lowering overall project expenses.

Adapting Technology for Affordability

Innovative technologies can be adapted or simplified to meet the needs of low-resource settings, ensuring affordability and accessibility.

• **Modular Systems**: Prefabricated components, such as modular green roof panels or bioswale units, can be deployed in low-cost configurations to reduce installation expenses.

• **Smart Tools for Maintenance**: Low-cost monitoring devices, such as rain gauges or basic water sensors, enable communities to track system performance without expensive technology.

• **DIY Solutions**: Promoting do-it-yourself (DIY) approaches empowers communities to construct and maintain green infrastructure using simple tools and locally sourced materials.

Policy Support and Institutional Innovation

Supportive policies and institutional frameworks are essential for enabling innovation in low-resource settings.

• **Simplified Regulations**: Streamlining permitting processes and reducing regulatory barriers makes it easier for communities to implement green infrastructure projects.

• **Subsidies and Incentives**: Governments can offer small grants or subsidies to encourage green infrastructure adoption. For example, subsidizing the cost of native plants or water-saving technologies makes these solutions more affordable.

• **Partnerships with NGOs**: Nonprofit organizations often play a vital role in supporting green infrastructure in low-resource settings by providing funding, expertise, and advocacy.

Scaling Through Knowledge Sharing

Sharing best practices and lessons learned from successful projects in similar contexts enables communities to replicate and scale green infrastructure innovations.

• **Peer Learning Networks**: Establishing networks among communities fosters the exchange of ideas, strategies, and resources for green infrastructure.

• **Case Studies and Toolkits**: Providing accessible case studies, guides, and toolkits equips stakeholders with practical knowledge to implement green solutions.

• **Capacity Building Programs**: Training programs conducted by NGOs or international organizations enhance technical skills and promote knowledge dissemination.

Building Financial Capacity Among Stakeholders

The successful implementation and sustainability of green infrastructure projects depend heavily on the financial capacity of stakeholders involved, including governments, private sector entities, nonprofits, and communities. Financial capacity encompasses the ability to secure funding, manage resources effectively, and ensure the long-term viability of green infrastructure initiatives. Strengthening this capacity enables stakeholders to overcome financial barriers and foster resilience in urban development.

Understanding Financial Capacity

Financial capacity refers to the ability of stakeholders to mobilize, allocate, and manage financial resources for green infrastructure projects. It includes skills and knowledge in budgeting, accessing funding mechanisms, assessing financial risks, and ensuring transparency in financial operations. A lack of financial capacity can hinder the planning, implementation, and maintenance of projects, particularly in resource-constrained settings.

Strategies for Building Financial Capacity

To enhance financial capacity among stakeholders, targeted strategies must be employed to address knowledge gaps, improve access to funding, and foster collaboration.

• **Financial Education and Training**: Providing stakeholders with training in financial planning, grant writing, and cost-benefit analysis equips them with the skills to secure and manage funding effectively. Workshops, online courses, and toolkits can address key topics, such as leveraging public-private partnerships or navigating innovative financing mechanisms.

• **Simplified Access to Funding**: Streamlining application processes for grants, loans, and subsidies ensures that stakeholders, particularly those with limited administrative resources, can access financial support for green infrastructure projects. Government programs and nonprofit organizations can play a pivotal role in simplifying access.

• **Resource Sharing and Collaboration**: Encouraging partnerships between stakeholders allows for resource pooling and knowledge sharing. For instance, a municipality might partner with local nonprofits or private firms to co-finance a green infrastructure project, leveraging shared expertise and funding.

• **Incentives for Participation**: Governments and financial institutions can incentivize stakeholders by offering tax benefits, fee reductions, or subsidies for adopting green infrastructure practices. These incentives reduce financial burdens and encourage investment.

The Role of Technology in Financial Capacity

Technology can enhance financial capacity by providing tools for resource management and performance tracking:

• **Digital Budgeting Tools**: Platforms that allow stakeholders to monitor budgets and allocate resources effectively streamline financial management.

• **Impact Measurement Systems**: Tools that quantify the benefits of green infrastructure, such as reduced stormwater runoff or increased biodiversity, help stakeholders demonstrate value and attract funding.

• **Crowdfunding Platforms**: Technology enables communities to mobilize financial resources from a broad audience, bridging funding gaps for small-scale projects.

Long-Term Sustainability

Building financial capacity is not a one-time effort but an ongoing process. Establishing mechanisms for continuous learning and adaptation ensures that stakeholders remain equipped to manage resources effectively over time. Regular assessments of financial practices, coupled with capacity-building programs, foster resilience and adaptability.

Chapter 9: Leveraging Technology for Green Infrastructure Financing

This chapter explores the various ways technology can be leveraged to overcome financial barriers, enhance resource efficiency, and attract diverse funding sources for green infrastructure. By embracing these technological innovations, cities and stakeholders can unlock new opportunities to finance sustainable urban development and drive the adoption of green infrastructure solutions worldwide.

Role of Digital Tools in Project Optimization

Digital tools are transforming how green infrastructure projects are planned, implemented, and managed, significantly enhancing their efficiency and impact. By providing data-driven insights, streamlining operations, and improving stakeholder coordination, these tools enable stakeholders to optimize project outcomes while reducing costs and resource use. The integration of digital technologies into green infrastructure planning is vital for addressing urban challenges such as stormwater management, urban heat islands, and flood risks.

Planning and Design Optimization

Digital tools play a critical role in the planning and design phase of green infrastructure projects, ensuring that resources are allocated effectively and solutions are tailored to local conditions.

• **GIS**: GIS enables precise mapping and spatial analysis, allowing stakeholders to identify areas most in need of green infrastructure interventions. For instance, GIS can highlight flood-prone zones, heat-affected neighborhoods, or areas with poor water quality.

• **Hydrological Modeling Software**: Tools such as SWMM (Storm Water Management Model) simulate the flow of stormwater through urban systems, helping designers predict the performance of green infrastructure solutions like bioswales or permeable pavements.

• **Building Information Modeling (BIM)**: BIM integrates data from various sources to create detailed 3D models of projects, enabling stakeholders to visualize designs, assess potential challenges, and refine plans before implementation.

These tools ensure that projects are strategically located, appropriately scaled, and designed to achieve maximum efficiency and impact.

Real-Time Monitoring and Performance Tracking

Once green infrastructure projects are implemented, digital tools are essential for monitoring their performance and ensuring they deliver the intended benefits.

• **Sensors and IoT Devices**: IoT devices provide real-time data on key performance metrics, such as stormwater retention, pollutant removal, and soil moisture levels. This data helps identify potential issues, such as blockages or underperformance, enabling timely interventions.

• **Remote Sensing**: Satellite imagery and drones allow for large-scale monitoring of green infrastructure, such as urban forests or restored wetlands. These tools provide insights into vegetation health, land use changes, and ecosystem impacts.

• **Data Dashboards**: Centralized platforms aggregate and visualize performance data, enabling stakeholders to track project outcomes and make data-driven decisions. For example, dashboards can display metrics like reduced stormwater runoff or improved water quality over time.

By offering continuous monitoring and feedback, digital tools help optimize green infrastructure performance and extend its lifespan.

Enhancing Resource Efficiency

Digital tools enable stakeholders to optimize resource use during project implementation and maintenance, reducing costs and environmental impacts.

• **Resource Allocation Software**: Tools like project management platforms help track budgets, schedules, and resource utilization, ensuring that projects stay on track and within budget.

• **Predictive Analytics**: Using historical and real-time data, predictive analytics tools forecast future challenges, such as increased rainfall intensity or urban heat, helping stakeholders plan proactive interventions.

• **Automated Systems**: Technologies like automated irrigation systems adjust water use based on real-time soil moisture levels, ensuring that resources are used efficiently while maintaining project health.

These tools allow for smarter resource allocation, reducing waste and maximizing the return on investment for green infrastructure projects.

Improving Stakeholder Collaboration

Collaboration among diverse stakeholders is essential for the success of green infrastructure projects. Digital tools facilitate communication, coordination, and transparency, fostering more effective partnerships.

• **Collaboration Platforms**: Tools like shared cloud platforms or project management software enable real-time collaboration between government agencies, private developers, and community groups. These platforms streamline workflows, document sharing, and decision-making processes.

• **Public Engagement Tools**: Apps and digital surveys engage communities in project planning and feedback, ensuring that green infrastructure addresses local needs and priorities.

• **Transparent Reporting Systems**: Digital tools that provide transparent updates on project progress and outcomes build trust among stakeholders, particularly investors and funding agencies.

Enhanced collaboration ensures that green infrastructure projects are well-supported, efficiently executed, and aligned with stakeholder goals.

Addressing Challenges with Digital Tools

While digital tools offer numerous benefits, their implementation can present challenges, particularly in resource-constrained settings. Common issues include:

• **Cost of Technology**: Acquiring and maintaining digital tools may require significant investment, which can be a barrier for smaller municipalities or communities.

• **Training and Capacity Building**: Stakeholders may need training to effectively use advanced tools, particularly in regions with limited technical expertise.

• **Data Security and Privacy**: As digital tools generate and store large amounts of data, ensuring its security and compliance with privacy regulations is essential.

Addressing these challenges requires investments in capacity building, subsidies for digital tools, and partnerships with technology providers to make tools more accessible and affordable.

Predictive Analytics and Investment Planning

Predictive analytics is transforming the way green infrastructure projects are planned and financed by enabling stakeholders to anticipate future challenges, evaluate risks, and optimize investment strategies. By leveraging historical data, real-time monitoring, and advanced modeling techniques, predictive analytics provides actionable insights that improve decision-making, enhance resource allocation, and attract funding for sustainable urban development. This technology has become a critical tool for investment planning, particularly in addressing climate-related risks and ensuring the long-term success of green infrastructure initiatives.

Understanding Predictive Analytics in Green Infrastructure

Predictive analytics involves using data and statistical algorithms to forecast future outcomes based on historical patterns and real-time inputs. In the context of green infrastructure, this technology helps stakeholders anticipate environmental, social, and financial impacts, allowing them to design projects that are resilient, cost-effective, and aligned with long-term sustainability goals.

Applications of predictive analytics in green infrastructure include:

• **Stormwater Management**: Forecasting rainfall patterns and stormwater flow to design systems capable of handling future extremes.

• **Flood Risk Reduction**: Identifying areas most vulnerable to flooding and modeling the effectiveness of green infrastructure solutions, such as wetlands and bioswales.

• **Urban Cooling**: Predicting the cooling effects of urban greening projects to prioritize interventions in heat-vulnerable neighborhoods.

Enhancing Investment Planning with Predictive Analytics

Investment planning for green infrastructure often involves assessing risks, estimating returns, and allocating resources efficiently. Predictive analytics supports these activities by providing data-driven insights that enhance the financial viability and strategic alignment of projects.

• **Risk Assessment**: Predictive models help identify and quantify risks, such as the likelihood of extreme weather events or infrastructure failure. These insights enable stakeholders to design projects that mitigate risks and reduce potential financial losses.

• **Return on Investment (ROI) Analysis**: By projecting long-term benefits, such as flood damage prevention or public health improvements, predictive analytics provides a clearer picture of a project's ROI. This information is crucial for attracting investors and securing funding.

• **Scenario Planning**: Stakeholders can use predictive tools to evaluate different investment scenarios, comparing the costs and benefits of various green infrastructure strategies under changing environmental and economic conditions.

Leveraging Predictive Tools for Financial Decision-Making

A range of predictive tools and technologies can be employed to improve investment planning for green infrastructure:

• **Hydrological Models**: Tools like SWMM (Storm Water Management Model) simulate stormwater runoff and drainage performance under different rainfall scenarios, helping stakeholders plan cost-effective stormwater systems.

• **Geospatial Analysis**: Predictive analytics combined with GIS enables stakeholders to identify high-priority areas for intervention based on environmental and socioeconomic factors.

• **Climate Projections**: Data from global and regional climate models informs long-term planning, ensuring that green infrastructure projects are resilient to future climate impacts.

• **Economic Models**: Predictive economic tools estimate the financial benefits of green infrastructure, such as property value increases, cost savings from reduced flood damage, and ecosystem service enhancements.

Attracting Investment Through Predictive Insights

Investors and funding agencies are more likely to support projects that demonstrate clear, data-backed benefits and risk mitigation strategies. Predictive analytics enhances transparency and builds confidence among investors by:

• **Providing Measurable Outcomes**: Predictive models generate quantifiable metrics, such as the volume of stormwater retained or the reduction in flood risks, which can be used to demonstrate project performance.

• **Improving Accountability**: Regular updates and forecasts based on predictive tools ensure that projects remain on track and deliver expected results, fostering trust among stakeholders.

• **Aligning with ESG Goals**: Predictive analytics supports ESG compliance by showcasing the environmental and social impacts of green infrastructure investments.

Overcoming Challenges in Predictive Analytics

Despite its potential, predictive analytics faces several challenges that must be addressed to maximize its effectiveness:

• **Data Availability**: Reliable and high-quality data is essential for accurate predictions. In some regions, data gaps or inconsistent records can limit the effectiveness of predictive models.

• **Technical Expertise**: Using advanced predictive tools requires technical skills and knowledge, which may not be readily available in all organizations or communities.

• **Integration with Existing Systems**: Predictive analytics must be integrated with other planning and financial systems to provide comprehensive insights and ensure seamless decision-making.

• **Cost of Implementation**: The initial investment in predictive tools and software can be high, creating barriers for smaller municipalities or resource-constrained stakeholders.

Addressing these challenges involves investments in capacity building, data infrastructure, and partnerships with technology providers.

Blockchain and Smart Contracts for Transparency

Blockchain technology and smart contracts are transforming the landscape of green infrastructure financing by enhancing transparency, accountability, and efficiency. These digital innovations provide secure, decentralized systems for tracking financial transactions, verifying project outcomes, and automating processes, making them particularly valuable for large-scale, multi-stakeholder projects. By leveraging blockchain and smart contracts, stakeholders can build trust, reduce administrative burdens, and attract diverse sources of funding.

Understanding Blockchain in Green Infrastructure Financing

Blockchain is a distributed ledger technology that records transactions in a secure and transparent manner. Each transaction is stored as a block in a chain, ensuring that data cannot be altered without consensus from the network. This immutability and transparency make blockchain an ideal tool for managing the financial and operational aspects of green infrastructure projects.

Key applications of blockchain in green infrastructure financing include:

• **Tracking Funds**: Blockchain allows stakeholders to monitor the flow of funds in real time, ensuring that resources are used as intended and reducing the risk of mismanagement or corruption.

• **Verifying Outcomes**: By integrating performance data into the blockchain, stakeholders can verify that green infrastructure projects meet agreed-upon outcomes, such as stormwater retention or urban cooling targets.

• **Facilitating Collaboration**: Blockchain provides a single, tamper-proof source of truth for all stakeholders, streamlining communication and coordination in complex projects.

Role of Smart Contracts

Smart contracts are self-executing agreements coded onto a blockchain. These contracts automatically trigger actions, such as payments, when predefined conditions are met. In green infrastructure financing, smart contracts offer several advantages:

• **Performance-Based Payments**: Payments can be automatically released when project outcomes, such as pollutant reduction or flood risk mitigation, are verified by sensors or third-party evaluators.

• **Efficiency and Cost Reduction**: Automating processes like disbursement and reporting reduces administrative costs and minimizes delays.

• **Accountability**: Smart contracts ensure that all parties fulfill their obligations, as payments and actions are contingent on meeting agreed-upon conditions.

For example, a city implementing a green roof initiative could use smart contracts to ensure that developers receive payments only after the roofs are inspected and meet performance criteria.

Enhancing Transparency with Blockchain

Transparency is critical for attracting investment and ensuring the success of green infrastructure projects. Blockchain technology provides several tools for improving transparency:

• **Real-Time Monitoring**: Financial transactions and project performance data are recorded on the blockchain in real time, allowing stakeholders to monitor progress and outcomes.

• **Immutable Records**: Blockchain's tamper-proof nature ensures that all data, from funding allocations to project milestones, is securely stored and verifiable.

• **Stakeholder Access**: Blockchain enables all stakeholders, including funders, governments, and communities, to access the same data, fostering trust and collaboration.

Applications in Green Infrastructure Projects

Blockchain and smart contracts are already being explored for use in green infrastructure financing:

• **Crowdfunding Platforms**: Blockchain-based platforms enable transparent, decentralized fundraising for green projects. Contributors can track how their funds are used and see the impact of their contributions.

• **Carbon Credits and Ecosystem Services**: Blockchain systems can track the generation and trading of carbon credits or payments for ecosystem services, ensuring that credits are not double-counted and benefits are fairly distributed.

• **Water Resource Management**: Smart contracts can automate water usage payments or incentives for projects that improve water quality or reduce consumption.

Challenges and Opportunities

While blockchain and smart contracts offer significant benefits, they also present challenges:

• **Technical Barriers**: Implementing blockchain systems requires specialized knowledge and infrastructure, which may be limited in some regions or organizations.

• **Cost of Adoption**: Initial setup costs for blockchain systems can be high, potentially deterring smaller projects or stakeholders with limited resources.

• **Regulatory Uncertainty**: The regulatory landscape for blockchain technology is still evolving, creating potential risks for stakeholders adopting the technology.

However, these challenges also create opportunities for partnerships and innovation. Collaboration with technology providers, government support for blockchain adoption, and capacity-building initiatives can help address barriers and accelerate adoption.

Future Innovations in Tech-Enabled Financing

The rapid advancement of technology is set to revolutionize how green infrastructure projects are financed, managed, and scaled. Future innovations in tech-enabled financing will provide stakeholders with powerful tools to address funding gaps, improve transparency, and optimize resource allocation. By leveraging cutting-edge technologies such as artificial intelligence (AI), decentralized finance (DeFi), and digital twins, cities can create more efficient, equitable, and scalable solutions for green infrastructure development.

Artificial Intelligence for Financial Decision-Making

AI has the potential to transform investment planning and resource management for green infrastructure. By analyzing vast amounts of data, AI algorithms can identify optimal financing strategies, assess project risks, and predict long-term benefits.

• **Risk Assessment**: AI can evaluate environmental, social, and financial risks associated with green infrastructure projects, enabling stakeholders to design resilient and cost-effective initiatives.

• **Investment Optimization**: Algorithms can recommend investment portfolios that balance financial returns with sustainability objectives, attracting socially responsible investors.

• **Predictive Maintenance**: AI-powered systems can predict maintenance needs for green infrastructure, reducing costs and ensuring long-term performance.

DeFi

Decentralized finance, or DeFi, uses blockchain technology to provide open and accessible financial services without traditional intermediaries. This innovation could democratize green infrastructure financing by creating new pathways for funding.

• **Tokenized Investments**: DeFi enables the creation of digital tokens representing stakes in green infrastructure projects, allowing individuals and organizations to invest directly and transparently.

• **Smart Lending Platforms**: Blockchain-based lending platforms could offer low-interest loans for sustainable projects, with repayment tied to performance metrics.

• **Crowdfunded Capital**: DeFi platforms facilitate global crowdfunding for green infrastructure, expanding access to diverse funding sources.

Digital Twin Technology

Digital twins, which are virtual replicas of physical assets, offer a new way to plan, monitor, and optimize green infrastructure projects. These digital models simulate real-world conditions, providing insights into performance and potential improvements.

• **Scenario Planning**: Stakeholders can use digital twins to test different financing and design scenarios, identifying the most cost-effective and impactful strategies.

• **Real-Time Monitoring**: Digital twins provide continuous data on project performance, enabling stakeholders to make data-driven adjustments.

• **Investor Engagement**: Interactive visualizations of green infrastructure projects help communicate their value to investors, fostering trust and support.

Integration with Global Sustainability Platforms

Future innovations will also focus on integrating green infrastructure financing with global sustainability platforms. These platforms could aggregate data, funding opportunities, and best practices, creating a centralized hub for stakeholders.

• **Standardized Reporting**: Unified platforms could streamline reporting on financial and environmental outcomes, simplifying compliance with international frameworks like the SDGs.

• **Collaborative Networks**: Global platforms foster partnerships between governments, investors, and communities, accelerating the adoption of innovative financing models.

Chapter 10: The Future of Green Infrastructure Financing

This chapter explores the trends and innovations that will define the future of green infrastructure financing. It examines the role of global policy frameworks, private sector involvement, and technological advancements in driving progress. Additionally, the chapter discusses the importance of equity and inclusivity, highlighting how future financing strategies must prioritize underserved communities to create truly sustainable and resilient urban environments.

Emerging Trends in Green Financing

As global priorities shift toward sustainability and climate resilience, the financing of green infrastructure is evolving to meet new demands. Emerging trends in green financing are characterized by innovative models, technological integration, and collaborative approaches that address financial barriers and unlock new opportunities. These trends are not only reshaping how green infrastructure projects are funded but also creating pathways for more equitable and inclusive urban development.

Expansion of Green Bonds and Climate Bonds

Green bonds and climate bonds have become prominent tools for financing sustainability projects, including green infrastructure. These instruments raise capital specifically for environmentally friendly initiatives, providing a reliable funding stream while attracting investors committed to sustainability goals.

• **Broader Accessibility**: The market for green bonds is expanding beyond developed countries, with emerging economies increasingly issuing bonds to fund green infrastructure projects.

• **Standardization and Certification**: Initiatives such as the Climate Bonds Standard provide clear criteria for certifying green bonds, ensuring transparency and attracting more investors.

• **Diverse Applications**: Green bonds are being used to finance a wide range of projects, from stormwater management systems to urban reforestation, enabling cities to scale green infrastructure effectively.

Integration of Performance-Based Financing

Performance-based financing models are gaining traction as they align financial returns with measurable environmental outcomes. By tying payments to results, these models incentivize innovation and accountability in green infrastructure projects.

• **Pay-for-Success Models**: Governments and private investors collaborate to fund projects, with payments contingent on achieving specific goals, such as reducing flood risks or improving water quality.

• **Outcome Verification**: The use of advanced monitoring tools ensures that performance metrics are met, building trust among stakeholders and attracting further investment.

• **Risk Sharing**: Performance-based financing distributes financial risks among stakeholders, encouraging private sector participation in green infrastructure initiatives.

Rise of DeFi

DeFi is emerging as a disruptive trend in green financing, leveraging blockchain technology to democratize access to funding and reduce reliance on traditional financial institutions.

• **Crowdfunding via Blockchain**: DeFi platforms enable communities to raise funds for green infrastructure projects through tokenized investments and decentralized crowdfunding campaigns.

• **Transparent Transactions**: Blockchain technology enhances transparency by providing a secure and immutable ledger of all financial transactions, ensuring accountability.

• **Global Participation**: DeFi platforms remove geographical barriers, allowing stakeholders worldwide to contribute to and benefit from green infrastructure projects.

Technological Advancements in Financing

The integration of advanced technologies is transforming how green infrastructure projects are planned, financed, and managed.

• **AI**: AI enhances investment decision-making by analyzing large datasets to predict project risks and returns, optimizing resource allocation.

• **Digital Twins**: Virtual replicas of physical assets help stakeholders visualize project outcomes and explore different financing scenarios, improving planning and investor confidence.

• **IoT**: IoT devices enable real-time monitoring of green infrastructure performance, ensuring projects meet financing conditions tied to measurable outcomes.

Greater Emphasis on Equity and Inclusion

Emerging trends in green financing increasingly focus on ensuring that the benefits of green infrastructure reach underserved and vulnerable communities.

• **Targeted Investments**: Funds are being directed toward projects in low-income neighborhoods to address environmental justice issues and improve urban equity.

• **Community-Based Financing**: Models that involve local stakeholders in funding decisions empower communities and ensure that green infrastructure projects align with their needs.

• **Blended Finance**: Combining public, private, and philanthropic resources enables projects in resource-constrained settings to access sufficient funding.

Policy Support and Public-Private Collaboration

Supportive policies and partnerships between governments and private sector entities are driving the adoption of innovative financing models.

• **Government Incentives**: Tax credits, grants, and subsidies encourage private investment in green infrastructure, reducing the financial burden on public agencies.

• **PPPs**: Collaborations between public and private stakeholders pool resources and expertise, enabling large-scale green infrastructure projects.

• **International Frameworks**: Global initiatives, such as the Paris Agreement and the SDGs, provide a policy foundation for green financing, fostering international collaboration.

Growth of Impact Investing

Impact investing continues to grow as investors increasingly prioritize financial returns alongside positive environmental and social outcomes.

• **Private Sector Engagement**: Corporations and institutional investors are integrating ESG criteria into their portfolios, channeling funds toward green infrastructure projects.

• **Blended Returns**: Impact investors seek projects that deliver measurable environmental benefits while providing financial returns, making green infrastructure an attractive asset class.

• **Investor Confidence**: Clear metrics and transparent reporting frameworks enhance confidence in the impact and viability of green infrastructure investments.

Integrating Equity and Inclusivity in Green Infrastructure Financing

Equity and inclusivity are essential components of sustainable green infrastructure financing. As cities address climate resilience and urban water management challenges, ensuring that green infrastructure projects benefit all communities—particularly those historically underserved or disproportionately affected by environmental risks—is critical. Integrating equity and inclusivity into financing strategies ensures that green infrastructure supports social well-being alongside environmental goals, creating more resilient and just urban systems.

Understanding Equity and Inclusivity in Green Infrastructure

Equity in green infrastructure financing involves addressing disparities in access to resources, benefits, and decision-making processes. Inclusivity ensures that all voices, particularly those from marginalized or low-income communities, are heard and valued in planning and implementation.

Key principles include:

• **Access to Benefits**: Ensuring that green infrastructure projects deliver tangible benefits, such as improved air quality, flood protection, and recreational spaces, to all neighborhoods, regardless of socioeconomic status.

• **Community Representation**: Engaging diverse stakeholders in project planning and decision-making to reflect the needs and priorities of all groups.

• **Addressing Historical Disparities**: Targeting investments to communities that have faced environmental neglect or systemic inequities.

Challenges in Achieving Equity and Inclusivity

Despite their importance, equity and inclusivity in green infrastructure financing face several challenges:

• **Funding Gaps**: Low-income neighborhoods often lack access to the financial resources needed to implement green infrastructure.

• **Disproportionate Impacts**: Vulnerable communities frequently bear the brunt of environmental risks, such as flooding or poor air quality, yet may not receive adequate support.

• **Exclusion from Decision-Making**: Communities most affected by environmental challenges are often underrepresented in planning processes, leading to solutions that may not address their needs.

Strategies for Integrating Equity and Inclusivity

To ensure that green infrastructure financing promotes equity and inclusivity, stakeholders can adopt the following strategies:

• **Community-Driven Financing Models**: Engaging communities in the design, funding, and implementation of green infrastructure ensures that projects align with local needs. For example, participatory budgeting allows residents to prioritize investments in their neighborhoods.

• **Targeted Investments**: Directing funding to underserved areas addresses environmental injustices and reduces disparities in access to green spaces and infrastructure benefits.

• **Capacity Building**: Providing education and training programs empowers community members to participate actively in green infrastructure projects, fostering local ownership and long-term sustainability.

• **Blended Finance**: Combining public, private, and philanthropic resources helps bridge funding gaps for communities with limited financial capacity, enabling equitable access to green infrastructure solutions.

• **Policy Frameworks for Equity**: Establishing policies that mandate equity-focused approaches in green infrastructure planning ensures that inclusivity is integrated into all stages of project development.

Case for Inclusive Financing Mechanisms

Inclusive financing mechanisms not only promote social equity but also enhance the overall success of green infrastructure projects:

• **Broader Support**: Engaging diverse stakeholders builds trust and support, increasing the likelihood of project acceptance and long-term maintenance.

• **Increased Effectiveness**: Solutions informed by local knowledge are more likely to address specific environmental and social challenges effectively.

• **Resilience and Stability**: Equitable distribution of green infrastructure benefits reduces vulnerabilities and strengthens community resilience to environmental risks.

Monitoring and Evaluating Equity Outcomes

To ensure accountability, stakeholders must establish metrics for measuring equity and inclusivity in green infrastructure financing:

• **Access Metrics**: Tracking the distribution of green infrastructure projects across different neighborhoods ensures fair allocation of resources.

• **Participation Metrics**: Measuring community involvement in planning and decision-making highlights inclusivity in the process.

• **Impact Metrics**: Evaluating the environmental, social, and economic benefits delivered to marginalized communities demonstrates progress toward equity goals.

Scaling Solutions for Global Sustainability

Scaling green infrastructure solutions globally is essential to addressing pressing environmental challenges, such as climate change, urbanization, and resource depletion. By expanding the adoption of sustainable practices across diverse regions and contexts, stakeholders can amplify the environmental, social, and economic benefits of green infrastructure. However, achieving global scalability requires innovative approaches, robust partnerships, and adaptable strategies that accommodate regional variations in resources, capacities, and needs.

Importance of Scaling Green Infrastructure

Green infrastructure offers proven benefits for urban water management, biodiversity enhancement, and climate resilience. Scaling these solutions globally ensures that their impact extends beyond isolated projects, contributing to broader sustainability goals, such as the United Nations SDGs. Key reasons for scaling include:

• **Addressing Global Challenges**: Climate risks, such as flooding, droughts, and heatwaves, require scalable solutions that can be implemented in diverse urban and rural settings.

• **Maximizing Co-Benefits**: Green infrastructure delivers multiple benefits, such as improved air quality, enhanced public health, and increased urban livability. Scaling these solutions allows more communities to access these advantages.

• **Advancing Equity**: Global scalability ensures that underserved and vulnerable populations, particularly in low-resource settings, also benefit from sustainable development initiatives.

Strategies for Scaling Green Infrastructure

Achieving scalability requires targeted strategies that foster collaboration, leverage technology, and adapt to local contexts.

• **Promoting International Collaboration**: Global partnerships between governments, international organizations, and private sector stakeholders enable the sharing of resources, knowledge, and expertise. For example, development banks and international NGOs can provide funding and technical assistance to scale green infrastructure in developing countries.

• **Leveraging Technology**: Advanced tools, such as GIS, remote sensing, and predictive analytics, streamline planning and implementation processes, making large-scale projects more feasible and efficient.

• **Adapting to Local Contexts**: Successful scaling requires tailoring green infrastructure solutions to regional environmental, cultural, and socioeconomic conditions. For instance, using native plants in rain gardens or designing stormwater systems that address specific local challenges ensures that projects are both effective and sustainable.

• **Building Local Capacity**: Providing education and training programs equips local stakeholders with the skills and knowledge needed to implement and maintain green infrastructure. This strategy fosters long-term sustainability and reduces reliance on external expertise.

Financing Global Scale

Expanding green infrastructure globally requires innovative financing mechanisms that attract diverse funding sources and support long-term investments.

• **Blended Finance**: Combining public, private, and philanthropic resources bridges funding gaps and ensures that projects are accessible to resource-constrained communities.

• **Green Bonds**: Issuing bonds tied to sustainable projects provides a scalable source of capital for large-scale green infrastructure initiatives.

• **Performance-Based Financing**: Linking financial returns to measurable outcomes ensures accountability and incentivizes high-impact projects.

Addressing Barriers to Scalability

While scaling green infrastructure offers significant potential, several challenges must be addressed to ensure success:

• **Resource Limitations**: Developing countries and underserved regions often face financial and technical constraints that hinder scalability. International support and capacity-building programs can help overcome these barriers.

• **Policy Fragmentation**: Inconsistent regulations across regions or countries can create obstacles to scaling. Aligning policies and standards with global sustainability frameworks, such as the SDGs, facilitates broader adoption.

• **Stakeholder Coordination**: Scaling requires collaboration among diverse stakeholders with varying priorities. Transparent communication and shared goals are essential to fostering effective partnerships.

Measuring and Monitoring Impact

To ensure the success of scaled solutions, robust monitoring and evaluation frameworks are needed. These frameworks provide data on project performance, environmental outcomes, and social benefits, enabling stakeholders to assess impact and refine strategies.

• **Global Metrics**: Standardized indicators, such as reductions in greenhouse gas emissions or improvements in water quality, allow stakeholders to measure progress consistently across regions.

• **Technology Integration**: Tools like IoT sensors and satellite imagery enable real-time tracking of green infrastructure performance, ensuring accountability and adaptive management.

• **Community Feedback**: Engaging local communities in monitoring processes ensures that projects address their needs and priorities effectively.

Examples of Scaled Green Infrastructure

Several initiatives demonstrate the potential for scaling green infrastructure globally:

• **China's Sponge Cities Program**: This large-scale initiative integrates green infrastructure into urban planning to improve water management and resilience in over 30 cities.

• **European Union's Green Infrastructure Strategy**: The EU promotes green infrastructure across member states, leveraging funding and policy frameworks to achieve regional sustainability goals.

• **Africa's Great Green Wall**: This ambitious project aims to combat desertification and restore ecosystems across the Sahel region through large-scale green infrastructure solutions.

Recommendations for Stakeholders

To advance green infrastructure financing and ensure its successful implementation, stakeholders across sectors must adopt strategies that align resources, foster collaboration, and address financial barriers. By tailoring their actions to their specific roles and responsibilities, governments, private sector actors, nonprofits, and communities can collectively create a robust framework for sustainable urban development.

Recommendations for Governments

Governments play a central role in setting policies, allocating resources, and driving the adoption of green infrastructure. Key recommendations include:

• **Develop Supportive Policies**: Establish clear regulations and incentives that encourage green infrastructure adoption, such as tax benefits, grants, and subsidies for sustainable projects.

• **Promote PPPs**: Facilitate collaborations with private entities to leverage resources and expertise for large-scale green infrastructure projects.

• **Align with Global Goals**: Integrate green infrastructure into national and local climate adaptation and mitigation plans, ensuring alignment with global frameworks like the SDGs and the Paris Agreement.

• **Invest in Capacity Building**: Provide technical training for municipal staff and local stakeholders to enhance their ability to plan, implement, and maintain green infrastructure projects.

Recommendations for Private Sector Stakeholders

The private sector's financial resources and innovation capabilities make it a key player in green infrastructure development. Recommendations include:

• **Invest in Sustainable Solutions**: Prioritize investments in green infrastructure projects that align with ESG criteria and demonstrate measurable environmental benefits.

• **Adopt Innovative Financing Models**: Explore performance-based financing, green bonds, and impact investing to support green infrastructure initiatives while achieving financial returns.

• **Collaborate with Public Entities**: Engage in PPPs and other partnerships to pool resources, share risks, and implement large-scale projects.

• **Leverage Technology**: Utilize tools like predictive analytics and blockchain to optimize project planning, monitor performance, and ensure transparency.

Recommendations for Nonprofit Organizations

Nonprofits and advocacy groups play a vital role in bridging gaps between stakeholders, advocating for equitable solutions, and providing technical support. Recommendations include:

• **Advocate for Equity and Inclusion**: Ensure that green infrastructure projects prioritize underserved communities, addressing environmental justice issues and promoting social equity.

• **Facilitate Community Engagement**: Act as intermediaries between governments, the private sector, and communities, ensuring that local voices are heard and integrated into project planning.

• **Provide Education and Resources**: Offer workshops, training, and toolkits to build local capacity and empower communities to participate in green infrastructure initiatives.

• **Secure Philanthropic Support**: Mobilize funding from foundations and donors to support green infrastructure projects in low-resource settings.

Recommendations for Communities

Communities are both beneficiaries and key participants in green infrastructure projects. Recommendations include:

• **Engage in Decision-Making**: Participate actively in planning processes to ensure that green infrastructure projects reflect local priorities and needs.

• **Support Maintenance Efforts**: Collaborate with local governments and nonprofits to maintain and monitor green infrastructure features, ensuring their long-term success.

• **Advocate for Funding**: Work collectively to secure local, regional, or national funding for projects, including through participatory budgeting or crowdfunding campaigns.

• **Educate Members**: Promote awareness of the benefits of green infrastructure within the community to build support and foster engagement.

Conclusion

Green infrastructure financing is a critical component of achieving sustainable urban development and addressing the pressing challenges of climate change, resource scarcity, and urbanization. Throughout this book, we have explored the innovative approaches, financial mechanisms, and collaborative strategies that are transforming the way cities manage water, enhance resilience, and promote environmental sustainability. By integrating green infrastructure into urban planning and policy, stakeholders can unlock significant environmental, social, and economic benefits while creating more equitable and livable urban spaces.

The future of green infrastructure financing depends on the collective efforts of governments, private sector actors, nonprofits, and communities. Each stakeholder plays a vital role in mobilizing resources, fostering innovation, and ensuring the long-term viability of green infrastructure projects. As the need for sustainable solutions grows, so does the opportunity to reimagine how cities address their water challenges and adapt to the demands of the 21st century.

In this conclusion, we recap the key themes discussed, highlight the urgency of investing in sustainable urban water solutions, and provide a call to action for stakeholders to embrace their roles in advancing green infrastructure financing.

Recap of Key Themes

This book has outlined the transformative potential of green infrastructure in urban water management, focusing on innovative financing models and collaborative strategies. We examined how tools like Environmental Impact Bonds, public funding mechanisms, and private sector engagement are driving investment in sustainable solutions such as rain gardens, permeable pavements, and green roofs. The integration of advanced technologies, such as blockchain and predictive analytics, enhances transparency and efficiency, making green infrastructure more accessible and impactful.

Additionally, the importance of equity and inclusivity has been emphasized, ensuring that all communities benefit from green infrastructure initiatives, particularly those historically underserved. Scaling solutions globally and aligning them with international frameworks like the Sustainable Development Goals are essential for addressing the challenges of climate change and urbanization. These themes highlight the interconnectedness of financial innovation, technological advancement, and social equity in creating a sustainable future.

Urgency of Sustainable Urban Water Investment

The growing threats of climate change, population growth, and aging infrastructure underscore the urgency of investing in sustainable urban water solutions. Cities worldwide are facing escalating challenges, including severe flooding, prolonged droughts, and declining water quality. Traditional gray infrastructure alone cannot address these complex issues, making green infrastructure a vital component of urban resilience.

Delaying investment in sustainable water management exacerbates environmental degradation, economic losses, and social inequalities. By acting now, stakeholders can reduce the long-term costs of climate impacts, enhance urban livability, and secure the well-being of future generations. Green infrastructure not only provides immediate benefits, such as stormwater management and urban cooling, but also supports broader sustainability goals, including biodiversity preservation and carbon sequestration. The time to act is now—proactive investment in green infrastructure is essential for building resilient cities that can thrive in the face of mounting environmental pressures.

Call to Action for Stakeholders

The successful implementation of green infrastructure financing depends on the collective commitment and collaboration of all stakeholders. Governments must lead by establishing supportive policies, allocating resources, and prioritizing green infrastructure in urban planning. Public-private partnerships should be leveraged to pool resources and expertise, enabling large-scale projects that deliver measurable environmental and social benefits.

The private sector must embrace innovative financing mechanisms, such as green bonds and performance-based contracts, to support green infrastructure while achieving financial returns. Nonprofits and advocacy groups should act as bridges, fostering community engagement and ensuring that equity and inclusivity remain at the forefront of green infrastructure initiatives.

Communities play a crucial role in supporting and sustaining these projects by participating in planning, monitoring outcomes, and advocating for local investments. Additionally, technology providers and researchers must continue to develop and refine tools that enhance efficiency, transparency, and scalability.

The challenges of climate change and urbanization demand bold, forward-thinking solutions. By embracing innovation, fostering collaboration, and committing to equity, stakeholders can transform urban water management and create sustainable, resilient cities. The journey toward a greener future starts with the collective actions of all who are invested in sustainability and the well-being of our planet.

www.ingramcontent.com/pod-product-compliance
Lightning Source LLC
Chambersburg PA
CBHW071557200326
41519CB00021BB/6799